The Interpretation of Ecological Data

The Interpretation of Ecological Data

A Primer on Classification and Ordination

E. C. Pielou
University of Lethbridge

A Wiley-Interscience Publication
JOHN WILEY & SONS
New York • Chichester • Brisbane • Toronto • Singapore

Library of Congress Cataloging in Publication Data:

Pielou, E. C., 1924–
The interpretation of ecological data.

"A Wiley-Interscience publication."
Bibliography: p.
Includes index.
1. Ecology—Statistical methods. 2. Multivariate analysis. I. Title.

QH541.15.S72P54 1984 574.5'072 84-7284
ISBN 0-471-88950-4

10 9 8 7 6

In Memory
of
Lydia

Preface

The aim of this book is to give a full, detailed, introductory account of the methods used by community ecologists to make large, unwieldy masses of multivariate field data comprehensible and interpretable. I am convinced that there is need for such a book. There are more advanced books that cover some of the same material, for example, L. Orlóci's *Multivariate Analysis in Vegetation Research* (W. Junk, 1978) and A. D. Gordon's *Classification. Methods for the Exploratory Analysis of Multivariate Data* (Chapman and Hall, 1981), but they assume a much higher level of mathematical ability in the reader than does this book. There are also more general discussions of the material, as in H. G. Gauch's *Multivariate Analysis in Community Ecology* (Cambridge University Press, 1982), or many of the chapters in the two volumes edited by R. H. Whittaker, *Ordination of Plant Communities* and *Classification of Plant Communities* (W. Junk, 1978), but they are more concerned with general principles, and with comparing the merits of different methods, than with explanations of the actual details of individual techniques.

Such explanations are sorely needed. Through this century, ecologists have used a series of aids to assist them in number-crunching, from tables of logarithms, through mechanical, then electrical, and then electronic desk calculators, to programmable computers. There has been no need for ecologists to understand how these aids work. But over the past decade, a new supply of "crutches" has appeared on the scene, namely, packaged computer programs. These enable ecologists to carry out elaborate analyses of their data without having to write their own programs. The packaged programs are often long and intricate, the work of computer experts. It would be unreasonable to demand that ecologists refrain from turning to these experts for help.

However, it is not unreasonable to expect ecologists to understand *what* the programs are doing for them even if they do not understand *how*. There is a world of difference between the person who uses a ready-made program to find the eigenvalues and eigenvectors of a large matrix and who understands what these things are, and the person who delegates the whole task of doing a principal component analysis (for instance) to such a program with no understanding of what the analysis does. Packaged programs are a mixed blessing. While they make it possible to analyze large bodies of data quickly, accurately, and in a way that best reveals their ecological implications, they also make it possible for inadequately trained people to go through the motions of data analysis uncomprehendingly.

This book is designed to help those who want to gain a complete understanding of the most popular techniques for analyzing multivariate data. It does not offer any computer programs. Instead, it demonstrates and explains the techniques using artificial data simple enough for all the steps in an analysis to be followed in detail from start to finish. The prerequisites are a knowledge of elementary algebra and coordinate geometry at about the first year undergraduate level. To make the book useful for self-instruction, exercises have been given at the ends of the chapters. The answers to them, and a comprehensive glossary, are at the end of the book.

I have written this book while holding an Alberta Oil Sands Technology and Research Authority (AOSTRA) Research Professorship at the University of Lethbridge. I am greatly indebted to the Authority, and the University, whose support has made the work possible. I also thank William Smienk of Lethbridge, who prepared all the figures.

E. C. PIELOU

Lethbridge, Alberta, Canada
April 1984

Contents

The Interpretation of Ecological Data

Chapter One

Introduction

Probably all ecologists are familiar with field notebooks whose pages look something like Figure 1.1. Probably most ecologists, even those still at the beginnings of their careers, have drawn up tables like it. Their efforts may be neater or messier, depending on the person and the circumstances (wind, rain, mosquitoes, gathering darkness, a rising tide, or any other of the stresses an ecologist is subject to). But such tables are all the same in principle. They show the values of each of several variables (e.g., species quantities) in each of several sampling units (e.g., quadrats). Tables such as these are the immediate raw material for community study and analysis.

Although natural, living communities as they are found in the field are, of course, an ecologist's ultimate raw material, it is impossible to come to grips with them mentally without first representing them symbolically. A table such as that in Figure 1.1, which is part of a *data matrix*,* is a typical symbolic representation of a natural community. It is the very first representation from which all subsequent analyses, and their representations, flow. Therefore, it is the first link in the chain leading from an actual, observed community to a theory concerning the community, and possibly to more inclusive theories concerning ecological communities generally.

The interpretation of such data matrices is the topic of this book. This introductory chapter first describes in general outline the procedures that make data interpretation possible. Then, as a necessary preliminary to all

*Words italicized in the text are defined in the Glossary at the end of the book.

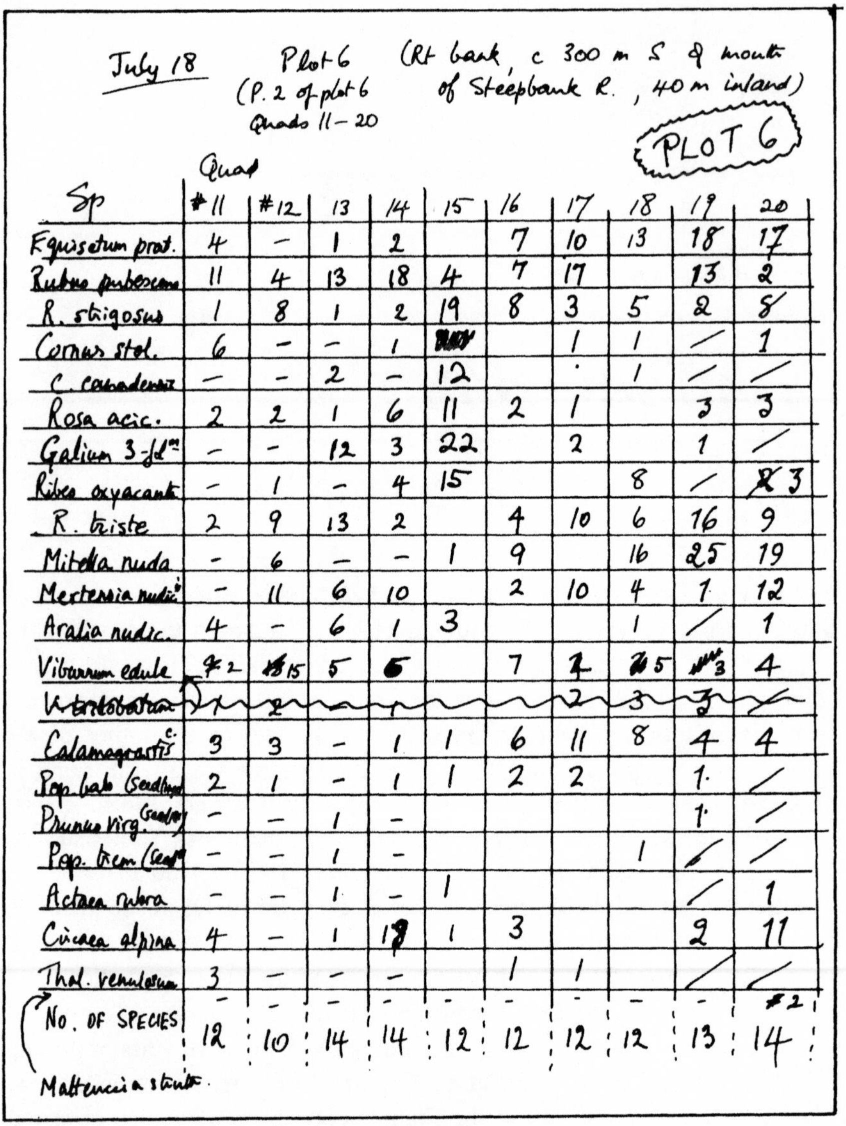

July 18 Plot 6 (Rt bank, c 300 m S of mouth of Steepbank R., 40 m inland)
(P. 2 of plot 6
Quads 11–20

PLOT 6

Sp / Quad	#11	#12	13	14	15	16	17	18	19	20
Equisetum prat.	4	–	1	2		7	10	13	18	17
Rubus pubescens	11	4	13	18	4	7	17		13	2
R. strigosus	1	8	1	2	19	8	3	5	2	8
Cornus stol.	6	–	–	1	[illegible]		1	1	/	1
C. canadensis	–	–	2	–	12		·	1	/	/
Rosa acic.	2	2	1	6	11	2	1		3	3
Galium 3-fld	–	–	12	3	22		2		1	/
Ribes oxyacanth.	–	1	–	4	15			8	/	~~2~~ 3
R. triste	2	9	13	2		4	10	6	16	9
Mitella nuda	–	6	–	–	1	9		16	25	19
Mertensia nudic.	–	11	6	10		2	10	4	1	12
Aralia nudic.	4	–	6	1	3			1	/	1
Viburnum edule	2	15	5	5		7	2	5	3	4
~~V. trilobatum~~	~~1~~	~~2~~		~~1~~			~~2~~	~~3~~	~~3~~	~~/~~
Calamagrostis c.	3	3	–	1	1	6	11	8	4	4
Pop. bals (seedling)	2	1	–	1	1	2	2		1	/
Prunus virg. (seedling)	–	–	1	–					1	/
Pop. trem (seedl.)	–	–	1	–				1	/	/
Actaea rubra	–	–	1	–	1				/	1
Circaea alpina	4	–	1	18	1	3			2	11
Thal. venulosum	3	–	–	–		1	1		/	/
No. of species	12	10	14	14	12	12	12	12	13	14 (#2)

Matteuccia struth.

Figure 1.1. A typical page from a field notebook. This one records observations on the ground vegetation in *Populus balsamifera* woodland in the floodplain of the Athabasca River, Alberta.

that follows, is a section on terminology. As is inevitable in a rapidly growing subject, a few of the technical terms are used in different senses by different writers. Therefore, it is necessary to define, unambiguously, the senses in which they are used in this book, as is done in Section 1.2.

1.1. DATA MATRICES AND SCATTER DIAGRAMS

A data matrix in the most general sense of the term is any table of observations made up of rows and columns. The data matrices most commonly encountered in community ecology are tables showing the amounts of several species in each of a number of sampling units. Thus there are obviously two possible ways of constructing a data matrix: either one may let each row represent a different species and each column a different sampling unit (as is done throughout this book), or vice versa. The method used here is the one favored by the majority of ecologists.

Any matrix (and that includes data matrices) can be denoted by a single symbol that represents, by itself, the whole array of numbers making up the table. A symbol representing a matrix is usually printed in boldface. If the matrix has s rows and n columns, it is described as an $s \times n$ matrix or, equivalently, as a matrix of order $s \times n$. The symbols s and n are used throughout this book to denote the orders of data matrices for mnemonic reasons: s stands for *s*pecies and n for *n*umber of sampling units. In specifying the order, or size, of a matrix, the number of rows is always written first and the number of columns second.

Now consider the symbolic representation of a 3×4 data matrix, say **X**, using subscripted xs in place of actual numerical values. Then

$$\mathbf{X} = \begin{pmatrix} x_{11} & x_{12} & x_{13} & x_{14} \\ x_{21} & x_{22} & x_{23} & x_{24} \\ x_{31} & x_{32} & x_{33} & x_{34} \end{pmatrix}.$$

As may be seen, every *element* of the matrix, that is, every individual term, has two subscripts. These subscripts specify the position of the element in the matrix: the first subscript gives the number of the row, and the second the number of the column in which the element occurs. For example, x_{24} denotes the element in the second row and fourth column of **X**; in general, one writes x_{ij} for the element in the ith row and jth column. This rule is adhered to universally in all mathematical writing where matrices appear. It follows that, in a data matrix, the jth column, say, constitutes a list of the species quantities in the jth sampling unit (if a species is absent from the unit its quantity is zero). Likewise, the ith row is a list of the quantities of species i in all the sampling units.

We now come to the problem of data interpretation, the subject of this book. It is seldom easy to perceive the latent structure in a "raw" data matrix as it is compiled in the field, or even to judge whether it has any structure. By "structure," we mean any systematic pattern that would indicate that, for example, certain groups of species tended to occur together, or that the sampling units, when appropriately arranged, would exhibit a gradual, continuous trend in their species compositions.

As an artificial example, consider the following 10×10 data matrix. It is a "raw" matrix.

$$\mathbf{X} = \begin{pmatrix} 2 & 2 & 0 & 0 & 3 & 0 & 1 & 1 & 4 & 3 \\ 3 & 0 & 0 & 0 & 0 & 0 & 4 & 0 & 1 & 2 \\ 0 & 0 & 4 & 3 & 0 & 2 & 0 & 1 & 0 & 0 \\ 1 & 3 & 0 & 0 & 4 & 1 & 0 & 2 & 3 & 2 \\ 0 & 1 & 3 & 4 & 0 & 3 & 0 & 2 & 0 & 0 \\ 0 & 4 & 0 & 1 & 3 & 2 & 0 & 3 & 2 & 1 \\ 0 & 2 & 2 & 3 & 1 & 4 & 0 & 3 & 0 & 0 \\ 3 & 1 & 0 & 0 & 2 & 0 & 2 & 0 & 3 & 4 \\ 4 & 0 & 0 & 0 & 1 & 0 & 3 & 0 & 2 & 3 \\ 0 & 3 & 1 & 2 & 2 & 3 & 0 & 4 & 1 & 0 \end{pmatrix};$$

as always, the rows represent species and the columns represent sampling units.

This matrix, undeniably, lacks any evident structure. But now suppose the sampling units (columns) and the species (rows) were rearranged in an appropriate fashion. The result is the "arranged matrix"

$$\mathbf{X} = \begin{pmatrix} 4 & 3 & 2 & 1 & 0 & 0 & 0 & 0 & 0 & 0 \\ 3 & 4 & 3 & 2 & 1 & 0 & 0 & 0 & 0 & 0 \\ 2 & 3 & 4 & 3 & 2 & 1 & 0 & 0 & 0 & 0 \\ 1 & 2 & 3 & 4 & 3 & 2 & 1 & 0 & 0 & 0 \\ 0 & 1 & 2 & 3 & 4 & 3 & 2 & 1 & 0 & 0 \\ 0 & 0 & 1 & 2 & 3 & 4 & 3 & 2 & 1 & 0 \\ 0 & 0 & 0 & 1 & 2 & 3 & 4 & 3 & 2 & 1 \\ 0 & 0 & 0 & 0 & 1 & 2 & 3 & 4 & 3 & 2 \\ 0 & 0 & 0 & 0 & 0 & 1 & 2 & 3 & 4 & 3 \\ 0 & 0 & 0 & 0 & 0 & 0 & 1 & 2 & 3 & 4 \end{pmatrix}.$$

This matrix contains exactly the same information as the earlier one. Only the orderings of the species, and of the sampling units, have been changed.

For instance, the species labeled #1 (therefore, appearing in the first row) in the raw matrix is labeled #4 (and, therefore, appears in the fourth row) in the arranged matrix. As to the columns, sampling unit #1, in column 1 of the raw matrix, for instance, has been relabeled as #2 and placed in column 2 in the arranged matrix.

The method by which the labeling system that produces the arranged matrix was determined is described in Chapter 4 and these two matrices, with their rows and columns labeled, are given again in Table 4.11. It suffices to remark here that the method is reasonably sophisticated. To derive the arranged matrix from the raw version without a prescribed method (an "algorithm") would be time-consuming and rather like the efforts of a noninitiate to solve a Rubik cube.

The preceding arranged matrix **X** is a clear example of a matrix with "structure." Data interpretation, as the term is used in this book, consists of methods of perceiving the structure in real data matrices even though these matrices, in raw form, may be as seemingly unstructured as the raw version of **X**. Table (or matrix) arrangement, as demonstrated here, is only one such method. Two other techniques, classification and ordination, do more to reveal the structure of a data matrix than does simple table arrangement, and descriptions of these techniques form the bulk of this book. Only a few, short introductory comments on these topics are necessary here.

Little need be said about classification. The word, as used in community ecology, has its ordinary everyday meaning (that of grouping similar things together into classes) and requires no special definition. Obviously, one can classify sampling units, uniting into classes those units that resemble one another in species composition; or one can classify species, uniting into classes those species that tend to occur together. More will be said later about these contrasted modes of classification.

Again, obviously, there are two contrasted ways in which one can go about the task of classifying data. If sampling units are to be classified, for example, one can either start with the individual units and combine them into small classes which are then themselves combined (and so on) thus successively forming ever larger classes; this is *agglomerative classification* (or *clustering*) and is the topic of Chapter 2. Or else one can start with the whole set of sampling units as the first, all-inclusive class and divide it into smaller classes; this is *divisive classification* and is the topic of Chapter 5.

We come now to *ordination*, a term that describes a whole battery of techniques that are useful in community ecology. In its original sense, the word means the same as "ordering." Consider, first, the data a plant ecologist might collect by observing a row of sample plots (*quadrats*) along an environmental gradient, for example up the side of a mountain or across a saltmarsh from land to sea. In this case the ordering of the quadrats is given in advance and there is no need for an "ordination." Now suppose the ecologist samples the herbaceous vegetation in level, mixed forest with randomly scattered quadrats, and suppose also that the environment, though moderately heterogeneous, exhibits no definite gradients. There would be no immediately obvious way to order the quadrats, but it might be reasonable to suppose that a "natural order" existed if only one could discover it. The supposition amounts to assuming that the environment of the forest consists of a mosaic of different habitats (not necessarily with clear boundaries between the mosaic patches) and that these different habitats have, themselves, a natural order in the same way that the successive habitats up a mountain, or across a saltmarsh, have a natural order. If this supposition is correct, then a technique can presumably be devised for discovering this natural order from the vegetation data. Such a technique yields an ordination.

The preceding paragraph describes the aim of the first practitioners of ecological ordination. Now let us approach the topic from a different starting point. To make the discussion concrete, visualize the data that would be amassed by a forest ecologist estimating the biomasses of the trees of several different species in a number of sampling plots in mixed forest. It is required to ordinate the plots. An obvious way of doing this would be to rank them according to the quantity they contained of the most abundant species. But why stop at one ordination? They could also be ranked according to the quantity they contained of the second most abundant species. An easy way to obtain these rankings would be to draw a scatter diagram in which the quantity of the first species is measured along the x-axis, and of the second species along the y-axis; each sample plot is represented by a point. Clearly, if the points were projected onto the x-axis, their order would correspond with that of the first ordination; likewise, if they were projected onto the y-axis, their order would correspond with that of the second ordination. But a clearer picture of community structure would be given by the scatter diagram itself; nothing is gained by considering the two axes separately. The scatter diagram can be considered as an ordination in two dimensions; it is a pictorial representation of the data, rather than merely a list of plot labels.

It should now be apparent where the argument is leading. If a two-dimensional scatter diagram (showing the amounts in the plots of the two most important species) is good, then a three-dimensional scatter diagram (showing the three most important species) is certainly better. Though harder to construct—it would have to be made with pins of various lengths stuck in corkboard—it would contain more information. But why stop at three species? Every time a species that is present is disregarded, a certain amount of information is sacrificed. Therefore, the best scatter diagram would be one displaying all the data, on all s species in the forest, regardless of how large s may be. The only difficulty is that a scatter diagram in more than three dimensions is impossible even to visualize, let alone construct. One is left with a "conceptual scatter diagram," an unsatisfactory object to investigate.

Ecological ordination as it is now practiced, however, does start from conceptual scatter diagrams. The various techniques amount to mapping these many-dimensional patterns in two dimensions (or, occasionally, three). In this way the unvisualizable conceptual patterns can be brought back into the real visible world to be looked at and studied.

The process is analogous to that of drawing two-dimensional maps of the three-dimensional globe. The difference is that when a geographer draws a map of the world on paper, he can easily consult a three-dimensional globe showing the true pattern of land and sea; the reduction in dimensionality is only from three to two. By contrast, an ecologist cannot see, or even visualize, the many-dimensional pattern that has to be mapped, and the required reduction in dimensionality is usually many times greater. All the same, the analogy between geographic mapping and ecological ordination is close, and it is also instructive. It shows that there are many possible techniques for mapping a many-dimensional scatter diagram in two dimensions, and that no one of them is automatically better than all the others. In the same way that different map projections are suited to different purposes, so different ordination techniques emphasize different aspects of the ecological data being examined. Unfortunately, however, the matching of technique to purpose is far less clear-cut in ecology than in geography; the purposes are not so well defined and methods for achieving them are not at all obvious.

We now turn to practical considerations. To arrive at a two-dimensional ordination of a many-dimensional scatter diagram, one has to operate on the given data, namely, an $s \times n$ data matrix (recall that s is the number of species and n the number of sampling units). The data may be thought of as giving the coordinates of the points in a scatter diagram plotted in a coordinate frame with s mutually perpendicular axes (one for each species), and consisting of *n data points* (one for each sampling unit). The coordinates of the *j*th point, say, are given by the elements in the *j*th column of the data matrix. The whole collection of data points, forming a "swarm" or "cloud," will be called a *data swarm*. All the many ordination techniques that ecologists use amount to different ways of mapping an *s*-dimensional data swarm on a sheet of two-dimensional paper. What is obtained is an ordination of sampling units. The best-known, most widely used methods are described in Chapter 4, following some necessary mathematical preliminaries in Chapter 3.

No doubt the reader has noticed that if it is legitimate to treat a body of data as equivalent to *n* points in *s*-dimensional space (*s*-space) as just described, then it is equally legitimate to treat it as *s* points in *n*-space. When this is done, each row (instead of each column) of the data matrix gives the coordinates of a data point. There are *s* points altogether and they are plotted in a coordinate frame with *n* mutually perpendicular axes, one for each sampling unit. Each point represents a species. Ordination of the data swarm, therefore, gives an ordination of species.

An ordination of sampling units is known as an *R-type ordination*, whereas an ordination of species is a *Q-type ordination*. Similarly, there are R-type and Q-type classifications, but these terms are seldom used.

The techniques for performing R-type and Q-type ordinations are identical, and at first thought the two types of analyses seem equally legitimate. Q-type analyses have one great drawback, however. If one plans to carry out any statistical tests on the data, it is essential that the "objects" sampled be independent of one another. Community sampling is nearly always done in a way that ensures the mutual independence of the sampling units, and the sampling units are the objects in an R-type ordination. But the species in the sampling units are almost certainly not independent and it is the species that are the objects in a Q-type ordination. Statistical hypothesis testing is outside the scope of this book, and we do not have occasion to consider the randomness and independence of sampling units again. However, the contrast between R and Q-type analyses from the statistical point of view should be kept in mind.

1.2. SOME DEFINITIONS AND OTHER PRELIMINARIES

Before proceeding it is most important to give the definitions used in this book of two terms much used in community ecology: *sample* and *clustering*.

The word *sample* is a source of enormous confusion in ecology, which is most unfortunate. To illustrate: imagine that a plant ecologist has made observations on a number of quadrats.* To a statistician, and to many ecologists, each quadrat is a *sampling unit* and the whole collection of quadrats is a *sample*. To other ecologists (e.g., Gauch, 1982) each individual quadrat is a "sample" and the whole collection of quadrats is a "sample set." The muddle can be shown most clearly in a 2 × 2 table in which the words in the four cells are the names given to the objects specified in the row labels by the people specified in the column labels. Thus:

	Statisticians and Some Ecologists	Other Ecologists
Single unit (e.g., quadrat)	Sampling unit	Sample
Collection of units	Sample	Sample set

This book uses the terms in the left-hand column. Neither terminology is entirely satisfactory, however, because it is a nuisance to have to use a two-word term (either sampling unit or sample set) to denote a single entity. Therefore, in this book I have used the word *quadrat* to mean a sampling unit of any kind, and have occasionally interpolated the additional words "or sampling unit" as a reminder. This is a convenient solution to the problem but it remains to be seen whether it will satisfy ecologists whose sampling units are emphatically not quadrats: for example, students of river fauna who collect their material with Surber samplers; palynologists, whose sampling unit is a sediment core; planktonologists whose sampling unit is the catch in a sampling net; entomologists, whose sampling unit is the catch in a sweep net or a light trap; diatom specialists, whose sampling unit is a microscope slide; foresters, whose sampling unit is a plot or stand that is much larger than a traditional quadrat although, like a quadrat, it is a delimited area of ground.

*The term *quadrat* is surely familiar to all ecologists. A. G. Tansley and T. F. Chipp, in their classic *Aims and Methods in the Study of Vegetation* (British Empire Vegetation Committee, 1926), define a quadrat as "simply a square area temporarily or permanently marked off as a sample of any vegetation it is desired to study closely." A more modern definition would omit the word "square"; a quadrat can be any shape. Note also that, although the definition says nothing about size, in ordinary usage a quadrat is thought of as smaller than a "plot." However, there is no agreed upon upper limit to the size of a quadrat, nor lower limit to the size of a plot; some "marked off areas" could reasonably be called by either name.

As far as possible I have avoided the word *sample* because of its ambiguity, but where it does appear, it is used in the statistical sense to mean a whole collection of quadrats.

The word *clustering* is also ambiguous. Some ecologists treat it as synonymous with *agglomerative classification*. Other ecologists treat it as synonymous with *classification* in the general sense, including both agglomerative and and divisive methods. Schematically, the two possibilities are as follows:

Classification	
Agglomerative (= *clustering*)	Divisive

Classification (= *clustering*)	
Agglomerative	Divisive

This book uses the scheme shown on the left.

It should also be remarked here that the word *cluster* is ambiguous too. Consider a two-dimensional swarm of data points. Some writers would apply the word *cluster* only to a "natural" (in other words, conspicuous) cluster of points, that is, a group of points that are simultaneously close to one another and far from all the rest. Other writers use the word *cluster* to mean any group of points assigned to the same class when a classification is done, no matter how arbitrary its boundaries. The word is not used in this book; hence there is no need to make a choice between these definitions.

A note on symbols should make subsequent chapters easier to follow. It was remarked that the term in the ith row and jth column of a data matrix [also known as the (i, j)th term, or as x_{ij}] denotes the quantity of species i in quadrat j. Throughout this book, the symbol i always represents a species and the symbol j a quadrat.

Matrices that are not data matrices are encountered in the book. Some of these show the relationship between pairs of species necessitating the use of two different symbols to represent two different species. In such cases the symbols are h and i.

Likewise, some matrices show the relationship between pairs of quadrats, necessitating the use of two different symbols to represent the two different quadrats. In such cases the symbols are j and k.

This convention should be recalled whenever paired subscripts are encountered. It is used in all but a few special contexts that are explained as they arise. Thus a term such as x_{ij}, with i and j as subscripts, relates to species i in quadrat j. A term such as y_{hi}, with h and i as subscripts, relates to a relationship between the two species h and i. And a term such as z_{jk}, with j and k as subscripts, relates to a relationship between quadrats j and k.

It has been remarked several times in this chapter that each element of a data matrix denotes the "amount" or "quantity" of a species in a quadrat. The way in which the quantity of a species should be measured depends, of course, on the kind of organism concerned.

For most animals, for most plankton organisms (plant or animal), for pollen grains, and for seedling plants of roughly equal size, the number of individuals is the simplest measure of quantity. The amounts of such organisms as mat-forming plants, and some colonial corals, sponges, and bryozoans, are often best measured as percentage "cover." The amounts of species in which, though the individuals are distinct, they are very unequal in size (such as the trees in uneven-aged forest) are best measured by the biomass of the individuals present in the quadrat.

In all community studies it is important to decide upon the best way to measure species quantities and then to make the measurements carefully. However, these matters are outside the purview of this book and are not referred to again.

1.3. AIM AND SCOPE OF THIS BOOK

The material covered in this book is listed in the Table of Contents. The aim of the book (to paraphrase what has already been said in the Preface) is to explain fully and in detail, at an elementary level, exactly how the techniques described actually work.

Packaged computer programs are readily available that can perform all these analyses quickly and painlessly. Too often users of these ready-made programs do no more than enter their data, select a few options, and accept the results in the printout with no comprehension of how the results were derived. But unless one understands a technique, one cannot intelligently judge the results.

Anyone who uses a ready-made program, for instance, to do a principal component analysis, should be capable of doing the identical analysis of a small, manageable, artificial data matrix entirely with a desk calculator, or if on a computer, then with programs written by oneself. Nobody can claim to understand a technique completely who is not capable of doing this.

It will be noticed that the book contains no mention of such topics as sampling errors, confidence intervals, and hypothesis tests. This is because the procedures described are treated as techniques for interpreting bodies of data that are interesting in their own right, and are not regarded merely as samples from some larger population. The techniques can safely be applied, as here described, as long as one realizes that what they reveal is the structure of the data actually in hand. A large body of data can certainly, by itself, provide rewarding ecological insights. But if it is intended to infer the structure of some parent population of which the data in hand are regarded only as a sample, then statistical matters do have to be considered. For example, it would probably be necessary, at the outset, to transform the observations so as to make their distribution multivariate normal. One is then entering the field of multivariate statistical analysis, which is wholly outside the scope of this book.

The distinction just made, between interpreting the patterns of given multidimensional data swarms, and analyzing multivariate statistical data, deserves emphasis; it is too often blurred. The two subjects are quite distinct and mastery of the first is a necessary prerequisite to appreciating the second. Students who wish to go on from the present book into the study of multivariate statistical analysis will find many books to choose from. Morrison (1976) and Tatsuoka (1971) can be especially recommended.

Chapter Two

Classification by Clustering

2.1. INTRODUCTION

The task described in this chapter is that of classifying, by clustering, a collection of sampling units. In all that follows, the sampling units are called *quadrats* for brevity and convenience. As described in Chapter 1, the data matrix has s rows, representing species, and n columns, representing quadrats. The (i, j)th element of the matrix represents the amount of species i (for $i = 1, \ldots, s$) in quadrat j (for $j = 1, \ldots, n$). We wish to classify the n quadrats by *clustering* or, as it is also called, by agglomeration.

To begin, each individual quadrat is treated as a cluster with only the one quadrat as member. As the first step, the two most similar clusters (i.e., quadrats) are united to form a two-member cluster. There are now $(n - 1)$ clusters, one with two members and all the rest still with only one member.

Next, the two most similar of these $(n - 1)$ clusters are united so that the total number of clusters becomes $(n - 2)$. The two clusters united may be single quadrats (one-member clusters), in which case two of the $(n - 2)$ clusters have two members and the rest one. Or else one of the two clusters united with another may be the two-member cluster previously formed; in that case one of the $(n - 2)$ clusters has three members and the rest one.

Again, the two most similar clusters are united. And again and again and again. The process continues until all the n original quadrats have been agglomerated into a single all-inclusive cluster.

Certain decisions need to be made before this process can be carried out. The questions to be answered are:

1. How shall the similarity (or its converse, the dissimilarity) between two individual quadrats be measured?
2. How shall the similarity between two clusters be measured when at least one and possibly both clusters have more than one member quadrat?

Both these questions can be answered in numerous ways. First, to answer question 1. Recall (see page 8) that, given n quadrats and s species, the data can be portrayed, conceptually, as n points (representing the quadrats) in an s-dimensional coordinate frame. Therefore, one possible way of measuring the dissimilarity between two quadrats is to use the *Euclidean distance*, in this s-space, between the points representing the quadrats. The coordinates of the jth of these n points are $(x_{1j}, x_{2j}, \ldots, x_{sj})$. This records the fact that quadrat j contains x_{1j} individuals* of species 1, x_{2j} individuals of species $2, \ldots,$ and x_{sj} individuals of species s.

The distance in s-dimensional space between the jth and kth points, denoted by $d(j, k)$, is, therefore,

$$d(j,k) = \sqrt{(x_{1j} - x_{1k})^2 + (x_{2j} - x_{2k})^2 + \cdots + (x_{sj} - x_{sk})^2}$$

$$= \sqrt{\sum_{i=1}^{s} (x_{ij} - x_{ik})^2}\,.$$

This is simply an extension to a space of s dimensions of the familiar result of Pythagoras's theorem whose two-dimensional version is shown in Figure 2.1.

The Euclidean distance between the points representing them is only one of the ways in which the dissimilarity of two quadrats might be measured. It is the measure we adopt in discussing four frequently used clustering techniques. Other ways of measuring dissimilarity are discussed in Section 2.6.

*Instead of measuring the amount of a species in a quadrat by counting the number of individuals, it is sometimes preferable to measure the biomass or, for many plant species, the areal "cover."

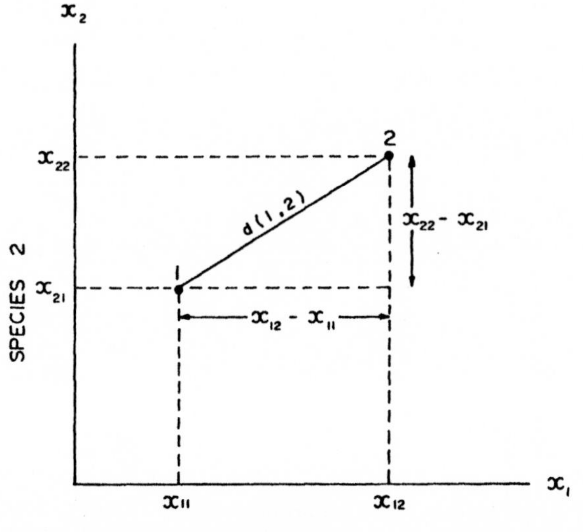

Figure 2.1. The distance between points 1 and 2, with coordinates (x_{11}, x_{21}) and (x_{12}, x_{22}), respectively, is $d(1,2)$. From Pythagoras's theorem,

$$d(1,2) = \sqrt{\{(x_{12} - x_{11})^2 + (x_{22} - x_{21})^2\}}.$$

Next for question 2, on how to measure the dissimilarity (now distance) between two clusters when each may contain more than one point (i.e., quadrat): the different ways in which this can be done are the distinguishing properties of the first three clustering methods described in the following.

2.2. NEAREST-NEIGHBOR CLUSTERING

In *nearest-neighbor clustering*, also known as *single-linkage clustering*, the distance between two clusters is taken to be the distance separating the closest pair of points such that one is in one cluster and the other in the other (see Figure 2.2).

Example. The following is a demonstration of the procedure applied to an artificially simple data matrix representing the amounts of 2 species in 10

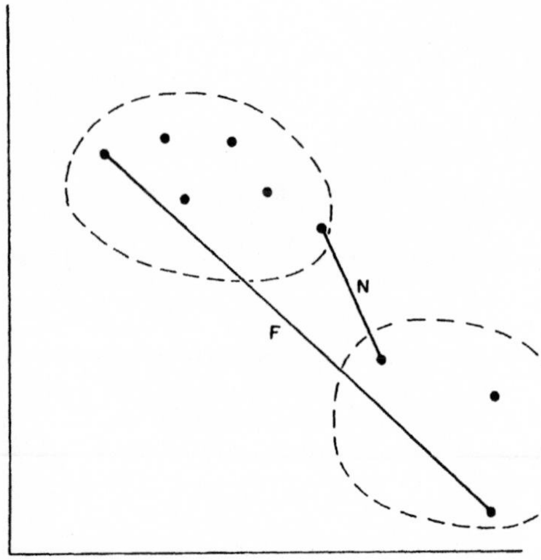

Figure 2.2. Two possible measures of the distance between two clusters. The nearest-neighbor distance N is the shortest distance, and the farthest-neighbor distance F is the longest distance between a member of one cluster and a member of the other.

quadrats. With only two species, it is possible to plot the data points in a plane and outline the successive clusters in the order in which they are created (see Figure 2.3).

Table 2.1 shows the data matrix (Data Matrix #1) and, below it, the *distance matrix*. In the distance matrix the numerical value of, for example, $d(3,5)$, the distance between the third and the fifth points, appears in the $(3,5)$th cell, which is the cell where the third row crosses the fifth column. It is $d(3,5) = 12.5$. Since the distance matrix must obviously be symmetrical, only its upper right half is shown.

The smallest distance in the matrix (in bold face type) is $d(5,8) = 2.2$. Therefore, the first cluster is formed by uniting quadrats 5 and 8; we call the cluster $[5,8]$.

The distance matrix is now reconstructed as shown in Table 2.2. In this new distance matrix distances to every point from the newly formed cluster $[5,8]$ are entered in row 5 and column 5; row 8 and column 8 are filled with asterisks to show that quadrat 8 no longer exists as a separate entity. The distance from $[5,8]$ to any point, for instance, to point 3, is the lesser of $d(3,5)$ and $d(3,8)$. Since $d(3,5) = 12.5$ and $d(3,8) = 14.4$, the distance

TABLE 2.1.

A. DATA MATRIX #1. THE QUANTITIES OF 2 SPECIES IN 10 QUADRATS.

Quadrat	1	2	3	4	5	6	7	8	9	10
Species 1	12	20	28	11	22	8	13	20	39	16
Species 2	30	18	26	5	15	34	24	14	34	11

B. THE DISTANCE MATRIX (THE ROW AND COLUMN LABELS ARE THE QUADRAT NUMBERS)

	Quadrat									
	1	*2*	*3*	*4*	*5*	*6*	*7*	*8*	*9*	*10*
1	0	14.4	16.5	25.0	18.0	5.7	6.1	17.9	27.3	19.4
2		0	11.3	15.8	3.6	20.0	9.2	4.0	24.8	8.1
3			0	27.0	12.5	21.5	15.1	14.4	13.6	19.2
4				0	14.9	29.2	19.1	12.7	40.3	7.8
5					0	23.6	12.7	**2.2**	25.5	7.2
6						0	11.2	23.3	31.0	24.4
7							0	12.2	27.9	13.3
8								0	27.6	5.0
9									0	32.5
10										0

TABLE 2.2. THE RECONSTRUCTED DISTANCE MATRIX AFTER THE FUSION OF QUADRATS 5 AND 8.

	1	*2*	*3*	*4*	*[5, 8]*	*6*	*7*	*8*	*9*	*10*
1	0	14.4	16.5	25.0	17.9	5.7	6.1	*	27.3	19.4
2		0	11.3	15.8	**3.6**	20.0	9.2	*	24.8	8.1
3			0	27.0	12.5	21.5	15.1	*	13.6	19.2
4				0	12.7	29.2	19.1	*	40.3	7.8
[5, 8]					0	23.3	12.2	*	25.5	5.0
6						0	11.2	*	31.0	24.4
7							0	*	27.9	13.3
8								0	*	*
9									0	32.5
10										0

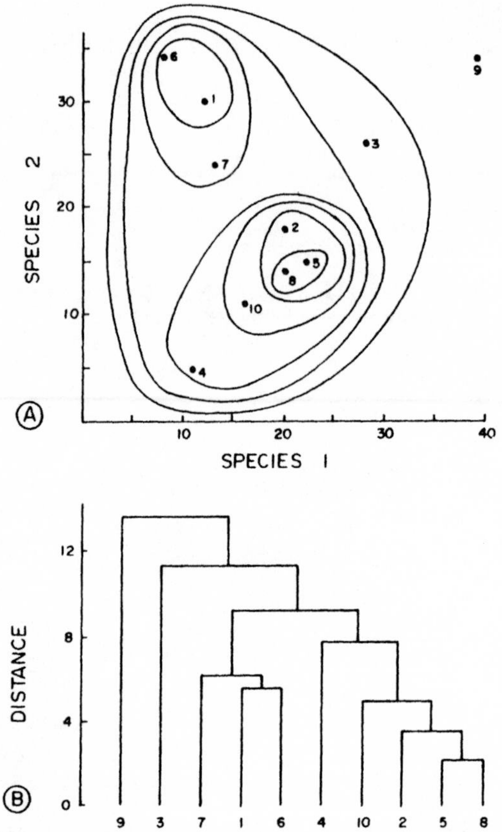

Figure 2.3. (*a*) The data points of Data Matrix #1 (see Table 2.1). The "contours" show the successive fusions with nearest-neighbor clustering except that, for clarity, the final contour enclosing all 10 points is omitted. (*b*) The corresponding dendrogram. Details are given in Table 2.3. The height of each node in the dendrogram is the distance between the pair of clusters whose fusion corresponds with the node.

from point 3 to the cluster [5, 8] is

$$d(3, [5,8]) = 12.5.$$

This values appears in the (3, 5)th cell of the reconstructed distance matrix. All the entries in the fifth row and column, which give the distances $d(j, [5,8])$ for all $j \neq 5$ or 8, are the lesser of $d(j,5)$ and $d(j,8)$.

The smallest entry in the reconstructed distance matrix is 3.6 (shown in boldface) in the (2, 5)th cell. Thus the next step in the clustering is the fusion

TABLE 2.3. STEPS IN THE NEAREST-NEIGHBOR CLUSTERING OF DATA MATRIX #1.

Step Number	Fusions[a]	Nearest Points[b]	Distance Between Clusters
1	5, 8	5, 8	2.2
2	[5, 8], 2	2, 5	3.6
3	[5, 8, 2], 10	8, 10	5.0
4	1, 6	1, 6	5.7
5	[1, 6], 7	1, 7	6.1
6	[5, 8, 2, 10], 4	4, 10	7.8
7	[5, 8, 2, 10, 4], [1, 6, 7]	2, 7	9.2
8	[5, 8, 2, 10, 4, 1, 6, 7], 3	2, 3	11.3
9	[5, 8, 2, 10, 4, 1, 6, 7, 3], 9	3, 9	13.6
10	All points are in one cluster		

[a]Unbracketed numbers refer to individual quadrats. The numbers in square brackets are the quadrats in a cluster.
[b]The distance between these two points defines the distance (given in the last column) between the two clusters united at this step.

of the existing two-member cluster [5, 8] with quadrat 2 to form the three-member cluster [2, 5, 8].

The distance matrix is reconstructed again, by adjusting the entries in row and column 2 and putting asterisks in row and column 5. And the procedure continues. The succession of steps is summarized in Table 2.3. At every step a new cluster is formed by the fusion of two previously existing clusters. (This includes one-member "clusters.") The final column in Table 2.3 shows the distance separating the clusters united at each step.

The procedure is shown graphically in Figure 2.3*a*; Figure 2.3*b* shows the result of the clustering as a *tree diagram* or *dendrogram*. The horizontal links in the dendrogram are known as *nodes* and the vertical lines as *internodes*. The height of each node above the base is set equal to the distance between the two clusters whose fusion the node represents. These distances are shown on the vertical scale on the left.

It should be noticed that the ordering of the points (quadrats) along the bottom is to some extent optional. Thus if the labels 1 and 6 were interchanged, or 5 and 8, there would be no change in the implications of the dendrogram. The dendrogram may be thought of as a hanging mobile with each node able to swivel freely where it is attached to the internode above it.

If we wished to divide the quadrats into classes, there are obviously several ways in which it could be done, all of them arbitrary. The arbitrariness arises because the points exhibit no natural clustering. The contours in Figure 2.3*a* do not represent abrupt discontinuities any more than the contour lines on a relief map of hilly country represent steps visible on the ground.

There are occasions, however, when an "unnatural" classification (sometimes called a dissection) is needed for practical purposes. For example, classification is required as a preliminary to vegetation mapping even if, in fact, the plant communities on the ground merge into one another with broad, indistinct ecotones. The lines separating communities on such a map are analogous to contour lines on a relief map, and are no less useful.

How to distinguish clusters, given a dendrogram like that in Figure 2.3*b*, is a matter of choice. Some common ways of classifying are as follows:

1. The number of classes to be recognized is decided beforehand. Thus suppose it had been decided to classify the 10 points in Data Matrix #1 into 4 classes. The memberships of the classes are found by drawing a horizontal line across the dendrogram at a level where it cuts four internodes. It will be seen that the resultant classes are [9], [3], [7, 1, 6], and [4, 10, 2, 5, 8].
2. The minimum distance that must separate clusters for them to be recognized as distinct may be decided beforehand. Suppose a minimum distance of 10 units were chosen in this example. Then three classes would be recognized, namely, [9], [3], and [7, 1, 6, 4, 10, 2, 5, 8].

If the internodes of a dendrogram are of conspicuously different lengths with short ones at the bottom and long ones at the top, then it follows that the points fall naturally into clusters without arbitrariness. Consider an example.

Example. Data Matrix #2 (see Table 2.4) shows the amounts of 3 species in 10 quadrats. The data points are shown graphically in Figure 2.4*a*, and the dendrogram resulting from nearest-neighbor clustering is in Figure 2.4*b*. The separation into three classes, namely, [1, 3, 2], [4, 5, 6, 7], and [8, 9, 10] is obvious in both diagrams, and no formal clustering procedure is needed to

TABLE 2.4. DATA MATRIX #2. THE QUANTITIES OF 3 SPECIES IN 10 QUADRATS.

Quadrat	1	2	3	4	5	6	7	8	9	10
Species 1	24	27	24	8	10	14	14	36	36	41
Species 2	32	30	29	20	18	20	22	14	9	12
Species 3	3	1	2	11	14	13	12	8	8	6

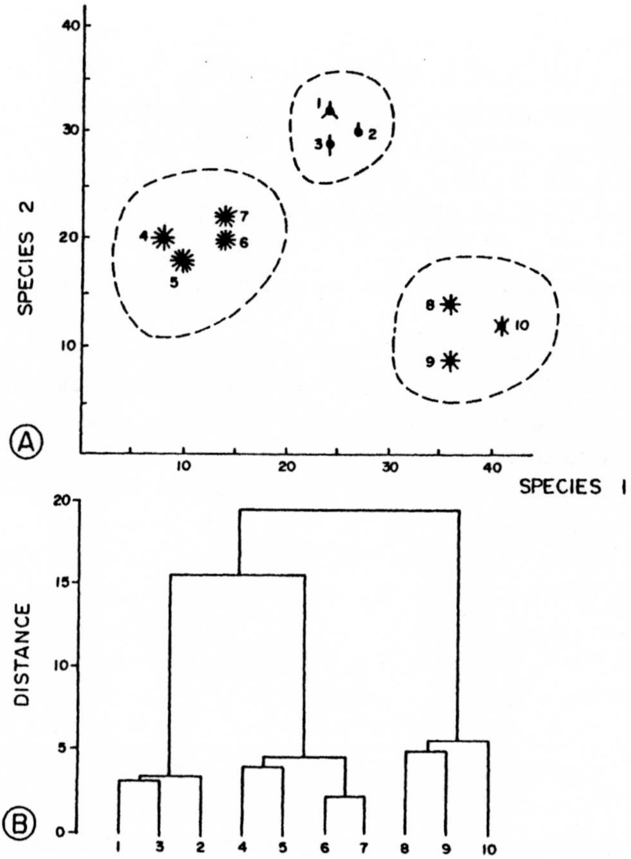

Figure 2.4. (*a*) The data points of Data Matrix #2 (see Table 2.4). The amount of species 3 in each quadrat is shown by the number of "spokes" attached to each point. (*b*) The dendrogram yielded by nearest-neighbor clustering.

recognize them. What makes the task easy, however, is the fact that, with only three species, the swarm of data points can be displayed in visualizable three-dimensional space, or by the device used in Figure 2.4*b* of representing the magnitude of the third coordinate by the number of "spokes" radiating from the points in a two-dimensional coordinate frame. When there are many species or, equivalently, when the swarm of data points occupies a many-dimensional space, visualization becomes impossible and a formal procedure is needed to produce a dendrogram.

Nearest-neighbor clustering is not often used in practice because it is prone to *chaining*. Chaining is the tendency for early formed clusters to grow by the accretion to them of single points one after another in succession. The effect can be seen in Figure 2.3 where the first, tight two-member cluster [5, 8] picks up the points 2, then 10, and then 4, one at a time, before uniting with another cluster containing more than one point. If a classification is intended to define classes for a purpose such as vegetation mapping, then a method that frequently leads to exaggerated chaining is defective. It results in clusters of very disparate sizes. Thus, as shown before, when the dendrogram in Figure 2.3*b* is used to define three clusters, two of them are "singletons" and all the remaining eight points are lumped together in the third cluster. For vegetation mapping, or for descriptive classifications generally, one usually prefers a method that yields clusters of roughly equal sizes. However, chaining may reveal a true relationship among the quadrats. Therefore, if what is wanted is a dendrogram that is in some sense "true to life," a clustering method that reveals natural chaining, if it exists, certainly has an advantage.

Indeed, a dendrogram is more than merely a diagram on which a classification can be based. It is a portrayal in two dimensions of a swarm of points occupying many dimensions (as many as there are species). A dendrogram need not be used to yield a classification. It can be studied in its own right as a representation of the interrelationships among a swarm of data points. Some workers find a dendrogram more informative than a two-dimensional ordination.

2.3. FARTHEST-NEIGHBOR CLUSTERING

In *farthest-neighbor clustering* (also known as *complete-linkage clustering*) the distance between two clusters is defined as the maximum distance between a point in one cluster and a point in the other (see Figure 2.2).

TABLE 2.5. STEPS IN THE FARTHEST-NEIGHBOR CLUSTERING OF DATA MATRIX #1.

Step Number	Fusions	"Farthest" Points[a]	Distance between Clusters
1	5, 8	5, 8	2.2
2	[5, 8], 2	2, 8	4.0
3	1, 6	1, 6	5.7
4	4, 10	4, 10	7.8
5	[1, 6], 7	6, 7	11.2
6	3, 9	3, 9	13.6
7	[5, 8, 2], [4, 10]	2, 4	15.8
8	[1, 6, 7], [5, 8, 2, 4, 10]	4, 6	29.2
9	[1, 6, 7, 5, 8, 2, 4, 10], [3, 9]	4, 9	40.3
10	All points are in one cluster		

[a] These are the quadrats whose distance apart, shown in the last column, defines the distance between the two clusters.

To apply the method to Data Matrix #1, we again start with the distance matrix in Table 2.1 and unite the two closest clusters (at this stage, individual points) which are, of course, points 5 and 8 as before. But in compiling each of the sequence of reconstructed distance matrices, we use the greater rather than the lesser of two distances. For example, the distance between cluster [5, 8] and point 2 is defined* as

$$d(2, [5,8]) = \max[d(2,5),\ d(2,8)] = \max(3.6, 4.0) = 4.0,$$

for farthest-neighbor clustering, whereas it was defined as

$$d(2, [5,8]) = \min[d(2,5),\ d(2,8)] = \min(3.6, 4.0) = 3.6$$

for nearest-neighbor clustering. The two clustering procedures (nearest-neighbor and farthest-neighbor) are the same in all respects except for this changed definition of intercluster distance. The result of clustering the data in Data Matrix #1 by farthest-neighbor clustering is shown in Table 2.5 and Figure 2.5. The figure should be compared with Figure 2.3. The difference is conspicuous.

Farthest-neighbor clustering has the "merit" that it tends to yield clusters that are fairly equal in size. This is because the farthest-neighbor distance

*$\max(x, y)$ denotes the maximum of x and y, and analogously for $\min(x, y)$.

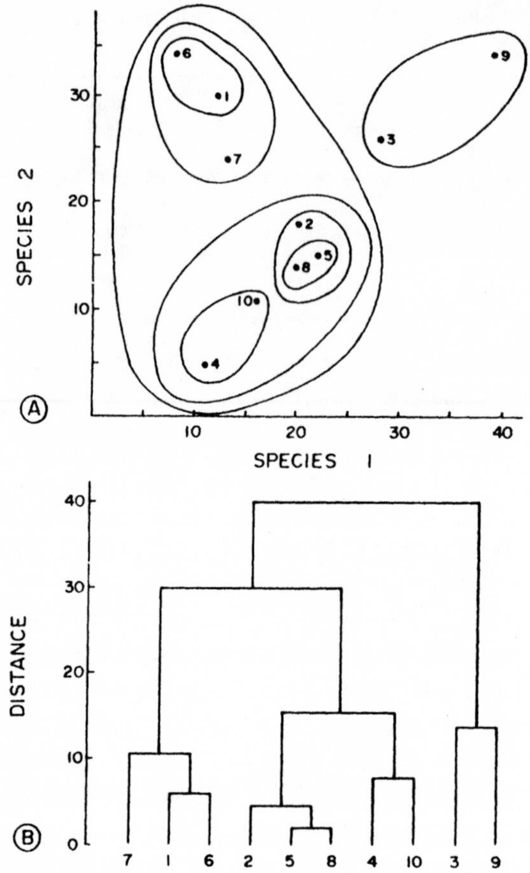

Figure 2.5. Farthest-neighbor clustering applied to Data Matrix #1. The dendrogram is based on Table 2.5.

between two populous neighboring clusters is often large in spite of the fact that, as a whole, they may be very similar. Consequently, it is more likely that an isolated unattached point at a moderate distance will be united with one of them than that they will unite with each other. Hence an anomalous quadrat may become a cluster member quite early in the clustering process and the fact that it is anomalous will be overlooked.

When true natural clusters exist, the outcomes of nearest-neighbor and farthest-neighbor clustering are usually very similar. Farthest-neighbor clustering applied to Data Matrix #2 gives results indistinguishable from those in Figure 2.4.

2.4. CENTROID CLUSTERING

Centroid clustering is one of several methods designed to strike a happy medium between the extremes of nearest-neighbor clustering on the one hand and farthest-neighbor clustering on the other. Nearest and farthest-neighbor methods have the defect that they are influenced at every step by the chance locations in the s-dimensional coordinate frame of only two individual points. That is, it is the distance between two points only that decides the outcome of each step. Centroid clustering overcomes this drawback by using a definition of intercluster distance that takes account of the locations of all the points in each cluster. To repeat, there are many ways in which this might be done, and centroid clustering, described in this section, is only one of the ways. A more general discussion of the various methods and how they are interrelated is given in Section 2.7.

In centroid clustering the distance between two clusters is taken to be the distance between their *centroids*. The centroid of a cluster is the point representing the "average quadrat" of the cluster; that is, it is a hypothetical quadrat containing the average quantity of each species, where the averaging is over all cluster members. Hence if there are m cluster members and s species, and if we write $(c_1, c_2, \ldots, c_s)$ for the coordinates of the centroid, then

$$c_1 = \frac{1}{m}(x_{11} + x_{12} + \cdots + x_{1m}) = \frac{1}{m}\sum_{j=1}^{m} x_{1j},$$

$$c_2 = \frac{1}{m}(x_{21} + x_{22} + \cdots + x_{2m}) = \frac{1}{m}\sum_{j=1}^{m} x_{2j},$$

and, in general,

$$c_i = \frac{1}{m}(x_{i1} + x_{i2} + \cdots + x_{im}) = \frac{1}{m}\sum_{j=1}^{m} x_{ij}. \tag{2.1}$$

For example, the centroid of the three-point cluster [2, 5, 8] in Data Matrix #1 has coordinates $[(20 + 22 + 20)/3, (18 + 15 + 14)/3] = (20.67, 15.67)$.

The clustering procedure is carried out in the same way as for nearest-neighbor and farthest-neighbor clustering. Thus each step consists in the fusion of the two nearest clusters (as before, a cluster may have only a single member); one finds which two clusters are nearest by searching the inter-cluster distance matrix for its smallest element. As in the methods already described, the distance matrix is reconstructed after each fusion by entering in it the distances from the newly formed cluster to every other cluster. But now the distances are those separating cluster centroids. The way in which the successive distance matrices are constructed is described after we have looked at some results.

The dendrogram produced by applying the centroid clustering procedure to Data Matrix #1 is shown in Figure 2.6. It should be noticed that the dendrogram is intermediate between that yielded by nearest-neighbor clustering (Figure 2.3*b*) and that yielded by farthest-neighbor clustering (Figure 2.5*b*). Thus in centroid clustering, as in farthest-neighbor clustering, points 3 and 9 unite to form the cluster [3, 9], whereas in nearest-neighbor clustering these two points are chained, one after the other, to the cluster formed by the remaining eight points. But centroid clustering, like nearest-neighbor clustering, chains point 10 and then point 4 to cluster [2, 5, 8], whereas in farthest-neighbor clustering, cluster [10, 4] is formed first and is only later united with [2, 5, 8].

Table 2.6, which resembles Tables 2.3 and 2.5, shows the steps in the centroid clustering of Data Matrix #1. Observe that instead of a column giving the distance between the two clusters united at each step, there is a column giving the square of this distance. The reason for this, together with

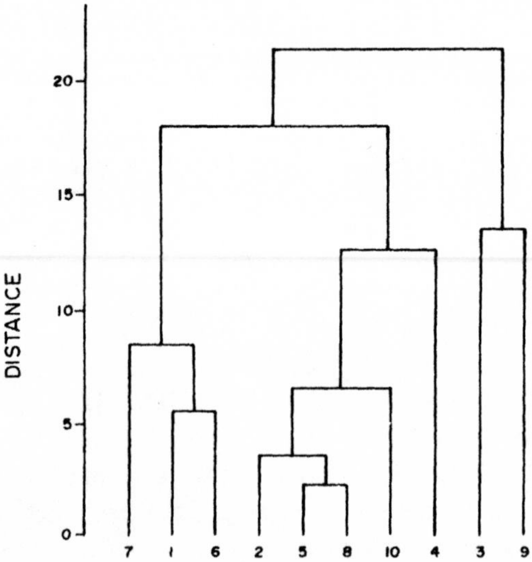

Figure 2.6. The dendrogram produced by applying centroid clustering to Data Matrix #1. It is based on Table 2.6.

TABLE 2.6. STEPS IN THE CENTROID CLUSTERING OF DATA MATRIX #1.

Step Number	Fusions	Square of Intercluster Distance
1	5, 8	5.00
2	2, [5, 8]	13.25
3	1, 6	32.00
4	10, [2, 5, 8]	43.56
5	7, [1, 6]	73.00
6	4, [2, 5, 8, 10]	162.50
7	3, 9	185.00
8	[7, 1, 6], [2, 5, 8, 10, 4]	326.25
9	[7, 1, 6, 2, 5, 8, 10, 4], [3, 9]	456.83
10	All points are in one cluster	

the derivation of these values, is explained in the following. Distances rather than distances-squared have been used to fix the heights of the nodes in the dendrogram so that the three dendrograms so far obtained from Data Matrix #1 may be easily compared.

We now consider the computations required in centroid clustering. Determining the distance between two cluster centroids is perfectly straightforward. Suppose there are s species, so that the distance required is that between two points in s-dimensional space or s-space. Let the two centroids be labeled C' and C''; their coordinates, obtained from Equation (2.1), are $(c_1', c_2', \ldots, c_s')$ and $(c_1'', c_2'', \ldots, c_s'')$, respectively. Then the square of the distance between C' and C'' is

$$d^2(C', C'') = (c_1' - c_1'')^2 + (c_2' - c_2'')^2 + \cdots + (c_s' - c_s'')^2.$$

As an example, recall Data Matrix #1 (see Table 2.1) and let C' and C'' be the clusters [7, 1, 6] and [2, 5, 8, 10, 4], respectively. The coordinates of C' are

$$(c_1', c_2') = \left[\tfrac{1}{3}(13 + 12 + 8), \tfrac{1}{3}(24 + 30 + 34)\right] = (11, 29.33);$$

analogous calculations show that the coordinates of C'' are

$$(c_1'', c_2'') = (17.8, 12.6).$$

Hence the square of the distance between them is

$$d^2(C',C'') = (c_1' - c_1'')^2 + (c_2' - c_2'')^2 = 326.24.$$

This corresponds (except for a rounding error in the final digit) with the squared intercluster distance opposite step #8 in Table 2.6.

Although this is, in principle, the most straightforward way of finding the square of the distance between two cluster centroids, it is inconvenient in practice. The inconvenience arises because the coordinates of the original data points have to be used in the calculations every time. It is more efficient computationally to derive the elements of each successive distance matrix from the elements of its predecessors. The following is a demonstration of the first few steps of the process applied to Data Matrix #1, after which the generalized version of the equation is given.

Each element in the initial distance matrix, more precisely a distance^2 matrix, is the square of the corresponding element in the distance matrix in Table 2.1. This initial matrix is at the top in Table 2.7. Its smallest element (in boldface) is $d^2(5,8) = 5.00$. Therefore, once again, step #1 is the formation of cluster [5, 8].

Now we do the first reconstruction of the distance^2 matrix. As before, since point 8 no longer exists as a separate entity, the elements of row and column 8 are replaced with asterisks. The fifth row and column, now labeled [5, 8], contain the distances-squared from the centroid of [5, 8] to every other point. The required distance-squared from the jth point is (as proved later)

$$d^2(j,[5,8]) = \tfrac{1}{2}d^2(j,5) + \tfrac{1}{2}d^2(j,8) - \tfrac{1}{4}d^2(5,8). \qquad (2.2)$$

Thus, when $j = 1$,

$$d^2(1,[5,8]) = \tfrac{1}{2}d^2(1,5) + \tfrac{1}{2}d^2(1,8) - \tfrac{1}{4}d^2(5,8)$$
$$= \tfrac{325}{2} + \tfrac{320}{2} - \tfrac{5}{4} = 321.25.$$

Likewise, when $j = 2$,

$$d^2(2,[5,8]) = \tfrac{13}{2} + \tfrac{16}{2} - \tfrac{5}{4} = 13.25,$$

and so on.

These values appear in the row and column labeled [5, 8] in the second distance^2 matrix in Table 2.7. All other elements remain as they were.

It is seen that the smallest element in the new matrix is 13.25 in row 2, column [5, 8]. Hence the second step in the clustering is the fusion of 2 and [5, 8] to form the three-member cluster [2, 5, 8].

TABLE 2.7. THE FIRST THREE DISTANCE² MATRICES CONSTRUCTED DURING CENTROID CLUSTERING OF DATA MATRIX #1.[a]

(1) The initial matrix. Each element is a distance^2 between two points.

	1	*2*	*3*	*4*	*5*	*6*	*7*	*8*	*9*	*10*
1	0	208	272	626	325	32	37	320	745	377
2		0	128	250	13	400	85	16	617	65
3			0	730	157	464	229	208	185	369
4				0	221	850	365	162	1625	61
5					0	557	162	5	650	52
6						0	125	544	961	593
7							0	149	776	178
8								0	761	25
9									0	1058
10										0

(2) The second matrix, after the fusion of points 5 and 8.

	1	*2*	*3*	*4*	*[5, 8]*	*6*	*7*	*8*	*9*	*10*
1	•	•	•	•	131.25	•	•	*	•	•
2		•	•	•	**13.25**	•	•	*	•	•
3			•	•	181.25	•	•	*	•	•
4				•	190.25	•	•	*	•	•
[5, 8]					•	549.25	154.25	*	704.25	37.25
6						•	•	*	•	•
7							•	*	•	•
8								•	*	*
9									•	•
10										•

(3) The third matrix, after the fusion of 2 and [5, 8].

	1	*[2, 5, 8]*	*3*	*4*	*5*	*6*	*7*	*8*	*9*	*10*
1	•	280.56	•	•	*	•	•	*	•	•
[2, 5, 8]		•	160.56	207.22	*	486.56	128.22	*	672.22	43.56
3			•	•	*	•	•	*	•	•
4				•	*	•	•	*	•	•
5					•	*	*	*	*	*
6						•	•	*	•	•
7							•	*	•	•
8								•	*	*
9									•	•
10										•

[a] In matrices 2 and 3 only elements differing from those in earlier matrices are shown. Unchanged elements are shown as dots.

The distance² matrix must now be reconstructed anew. The elements of row and column [5, 8] are replaced with asterisks. The new elements for the second row and column, now labeled [2, 5, 8], are found from the formula

$$d^2(j, [2,5,8]) = \tfrac{1}{3}d^2(j,2) + \tfrac{2}{3}d^2(j, [5,8]) - \tfrac{2}{9}d^2(2, [5,8]). \quad (2.3)$$

When $j = 1$,

$$d^2(1, [2,5,8]) = \tfrac{208}{3} + \tfrac{2}{3} \times 321.25 - \tfrac{2}{9} \times 13.25 = 280.56;$$

when $j = 3$,

$$d^2(3, [2,5,8]) = \tfrac{128}{3} + \tfrac{2}{3} \times 181.25 - \tfrac{2}{9} \times 13.25 = 160.56;$$

and so on. These values appear in the row and column labeled [2, 5, 8] in the third matrix in Table 2.7.

Equations (2.2) and (2.3) are particular examples of a general equation which we now derive. It is the equation for the distance² from any point (or cluster centroid) P to the centroid Q of an $(m + n)$-member cluster created by the fusion of two clusters $[M_1, M_2, \ldots, M_m]$ and $[N_1, N_2, \ldots, N_n]$ with m and n members, respectively. The centroids of these clusters are M and N.

The set-up is shown in Figure 2.7.

$$\text{Let } MP = a,\ NP = b, \text{ and } MN = c.$$

$$\text{Let the angles } \widehat{MQP} = \alpha \text{ and } \widehat{NQP} = \beta \text{ with } \alpha + \beta = 180°.$$

The distance required is x^2, where $PQ = x$. Recall that x^2 is needed as an element of the row or column headed $[M_1, M_2, \ldots, M_m, N_1, N_2, \ldots, N_n]$ in a distance² matrix undergoing reconstruction as one of the steps in a clustering operation. The values of a^2, b^2, and c^2 are known since they are elements in the distance² matrix constructed at an earlier step.

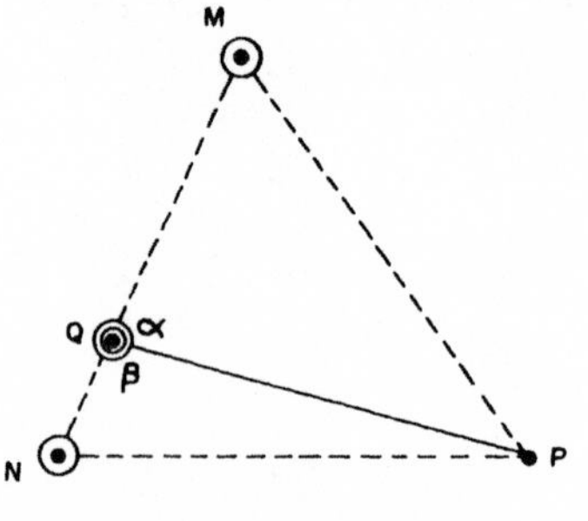

Figure 2.7. Illustration of the derivation of Equation (2.7). Q is the centroid of clusters M and N; it is assumed that $n > m$ and, therefore, Q is closer to N than to M. See text for further details.

As a preliminary to finding x^2 it is necessary to find MQ (or $NQ = c - MQ$). Since M and N are, respectively, the centroids of m-member and n-member clusters and Q is at their center of gravity, it is clear that

$$MQ = \frac{nc}{m+n} \quad \text{and} \quad NQ = \frac{mc}{m+n}.$$

Now, from Apollonius's theorem,*

$$(PQ)^2 + (MQ)^2 - 2(PQ)(MQ)\cos\alpha = (MP)^2$$

and

$$(PQ)^2 + (NQ)^2 - 2(PQ)(NQ)\cos\beta = (NP)^2.$$

Equivalently,

$$x^2 + \frac{n^2c^2}{(m+n)^2} - \frac{2xnc}{m+n}\cos\alpha = a^2; \tag{2.4}$$

$$x^2 + \frac{m^2c^2}{(m+n)^2} - \frac{2xmc}{m+n}\cos\beta = b^2. \tag{2.5}$$

Multiply (2.5) by n/m, and do the substitution $\cos\beta = \cos(180° - \alpha) = -\cos\alpha$. Then

$$\frac{nx^2}{m} + \frac{nmc^2}{(m+n)^2} + \frac{2xnc}{m+n}\cos\alpha = \frac{b^2n}{m}. \tag{2.6}$$

Add (2.4) and (2.6) to eliminate $\cos\alpha$. The sum is

$$x^2\left(1 + \frac{n}{m}\right) + \frac{nc^2}{(m+n)^2}\cdot(m+n) = a^2 + \frac{b^2n}{m}$$

whence

$$x^2\left(\frac{m+n}{m}\right) + \frac{nc^2}{m+n} = \frac{a^2m + b^2n}{m}.$$

*A proof of Apollonius's theorem is given as an appendix to this chapter.

Multiply through by $m/(m+n)$ to obtain

$$x^2 + \frac{mnc^2}{(m+n)^2} = \frac{a^2m + b^2n}{m+n}$$

or

$$x^2 = \frac{m}{m+n}a^2 + \frac{n}{m+n}b^2 - \frac{mn}{(m+n)^2}c^2. \qquad (2.7)$$

It is seen that (2.7) is the required general form of (2.2) and (2.3).

It is now apparent that when centroid clustering is being done, the most convenient measure of the dissimilarity between two clusters is the square of the distance separating their centroids rather than the distance itself (but see page 46). The terms x^2, a^2, b^2, and c^2 in Equation (2.7) are all squared distances, and use of this equation makes it easy to construct each distance^2 matrix from its predecessor as clustering proceeds from step to step. There is no simple relation among x, a, b, and c when they are not squared.

Centroid clustering has been described with greater mathematical rigor by Orloci (1978), who gives an example. He uses the name "average linkage clustering" for the method. Average linkage clustering is an inclusive term for several similar clustering procedures of which centroid clustering is one. The interrelationships among the several methods are briefly described in Section 2.7.

2.5. MINIMUM VARIANCE CLUSTERING

This is the last clustering method that is fully described in this book. Before going into details, it is necessary to define the term *within-cluster dispersion*, and to give two methods for computing it. The first of these methods is the obvious one, implied by the definition. The second, nonobvious method is a way of obtaining the identical result by a computationally simpler route.

First, the definition: the within-cluster dispersion of a cluster of points is defined as the sum of the squares of the distances between every point and the centroid of the cluster.

Next, we illustrate the computations.

Example. Consider Data Matrix #3 shown in Table 2.8. It lists the quantities of each of two species in five quadrats; it can be represented

TABLE 2.8. DATA MATRIX #3. THE QUANTITIES OF TWO SPECIES IN FIVE QUADRATS.

Quadrat	1	2	3	4	5
Species 1	11	36	16	8	28
Species 2	14	30	20	12	32

graphically by a swarm of five points (representing the quadrats) in a space of two dimensions, that is, a two-dimensional coordinate frame with axes representing the species.

Suppose the five points have been combined into a single cluster, as they will have been when the last step in a clustering process is complete. The centroid of the five-point cluster has coordinates

$$(c_1, c_2) = \left(\frac{11 + 36 + 16 + 8 + 28}{5}, \frac{14 + 30 + 20 + 12 + 32}{5} \right)$$
$$= (19.8, 21.6).$$

Now write $Q[1,2,3,4,5]$ for the within-cluster dispersion of the cluster of points $1, 2, 3, 4$ and 5; let $d^2(j, C)$ be the square of the distance from the centroid to the jth point.

Then, from the definition,

$$\begin{aligned} Q[1,2,3,4,5] &= \sum_{j=1}^{5} d^2(j,C) \\ &= d^2(1,C) + d^2(2,C) + \cdots + d^2(5,C) \\ &= \left[(11 - 19.8)^2 + (14 - 21.6)^2\right] \\ &\quad + \left[(36 - 19.8)^2 + (30 - 21.6)^2\right] \\ &\quad + \cdots + \left[(28 - 19.8)^2 + (32 - 21.6)^2\right] \\ &= 892. \end{aligned}$$

A simpler way of obtaining the same result is to use the equation

$$Q[1,2,3,4,5] = \frac{1}{n} \sum_{j<k} d^2(j,k).$$

(For a proof, see Pielou, 1977, p. 320.) Here $n = 5$, the number of points in the cluster; $d^2(j,k)$ is the squared distance between points j and k; the summation is over every possible pair of points, taking each pair once. This is the reason for putting the condition $j < k$ below the summation sign. It ensures that, for example, $d^2(1,2)$ shall be a component of the sum, but not $d^2(2,1)$ which is merely a repetition of $d^2(1,2)$. There are $n(n-1)/2 = 10$ distinct pairs of points, and hence 10 distinct components of the form $d^2(j,k)$. Thus

$$Q[1,2,3,4,5] = \tfrac{1}{5}\{d^2(1,2) + d^2(1,3) + \cdots + d^2(4,5)\}$$

with 10 components between the braces. Now

$$d^2(1,2) = (11-36)^2 + (14-30)^2 = 881,$$
$$d^2(1,3) = (11-16)^2 + (14-20)^2 = 61,$$
$$\cdots\cdots\cdots\cdots\cdots\cdots\cdots\cdots\cdots\cdots$$
$$d^2(4,5) = (8-28)^2 + (12-32)^2 = 800.$$

Hence

$$Q[1,2,3,4,5] = \tfrac{1}{5} \times 4460 = 892,$$

as before.

For a two-member cluster, say of points a and b, the within-cluster dispersion is

$$Q[a,b] = \tfrac{1}{2}d^2(a,b);$$

that is, it is half the square of the distance between them.

For a one-member cluster, say point a by itself, the within-cluster dispersion is zero; that is, $Q[a] = 0$.

We are now in a position to describe minimum variance clustering. At each step, those two clusters are to be united whose fusion yields the least increase in within-cluster dispersion. It is important to notice that what matters is not simply the value of the within-cluster dispersion of a newly formed cluster, but the amount by which this value exceeds the sum of within-cluster dispersions of the two separate clusters whose fusion formed the new cluster.

For example, consider the two clusters $[a, b, c]$ and $[d, e]$, having within-cluster dispersions of $Q[a, b, c]$ and $Q[d, e]$, respectively. Suppose they are united to form a new cluster whose within-cluster dispersion is $Q[a, b, c, d, e]$. The increase in within-cluster dispersion that this fusion has brought about, denoted by $q([a, b, c], [d, e])$, is

$$q([a,b,c],[d,e]) = Q[a,b,c,d,e] - Q[a,b,c] - Q[d,e].$$

It is values such as $q([a, b, c], [d, e])$ that are the criteria for deciding which two clusters should be united at each step of the clustering process. At every step the clusters to be united are always the two for which the value of q is least.

As with the clustering procedures described in earlier sections, the minimum variance method also requires the construction of a sequence of "criterion" matrices. Then the position in the matrix of the numerically smallest element indicates which clusters are to be united next. The matrices obtained when minimum variance clustering is applied to Data Matrix #3 are shown in Table 2.9. In addition to the sequence of criterion matrices $\mathbf{Q}_1$, $\mathbf{Q}_2$, $\mathbf{Q}_3$, and $\mathbf{Q}_4$, and printed above them, is the matrix $\mathbf{D}^2$. It is a distance^2 matrix whose elements are the squares of the distances separating every pair of points. The elements of $\mathbf{D}^2$ are used to construct the successive criterion matrices.

We now carry out minimum variance clustering on the data in Data Matrix #3. $\mathbf{Q}_1$, the first criterion matrix, has as its elements the within-cluster dispersion of every possible two-member cluster that could be formed by uniting two individual points. It has already been shown that for a two-member cluster consisting of points j and k, say, the within-cluster dispersion is

$$Q[j,k] = \tfrac{1}{2}d^2(j,k).$$

Therefore, the elements of $\mathbf{Q}_1$ are simply one-half the values of the corresponding elements of $\mathbf{D}^2$.

The smallest element in $\mathbf{Q}_1$ is 6.5 (shown in boldface) in cell (1, 4). Therefore, the first cluster to be formed is [1, 4]. We now construct the next criterion matrix $\mathbf{Q}_2$. It has asterisks in row and column 4, since point 4 no

TABLE 2.9. SUCCESSIVE MATRICES CONSTRUCTED IN THE MINIMUM VARIANCE CLUSTERING OF DATA MATRIX #3.

The Distance^2 Matrix

		1	*2*	*3*	*4*	*5*
	1	0	881	61	13	613
	2		0	500	1108	68
$\mathbf{D}^2 =$	*3*			0	128	288
	4				0	800
	5					0

The Sequence of Criterion Matrices

		1	*2*	*3*	*4*	*5*
	1	0	440.5	30.5	**6.5**	306.5
	2		0	250	554	34
$\mathbf{Q}_1 =$	*3*			0	64	144
	4				0	400
	5					0

		[*1*, *4*]	*2*	*3*	*4*	*5*
	[*1*, *4*]	0	660.83	60.83	*	468.83
	2		0	250	*	**34**
$\mathbf{Q}_2 =$	*3*			0	*	144
	4				0	*
	5					0

		[*1*, *4*]	[*2*, *5*]	*3*	*4*	*5*
	[*1*, *4*]	0	830.25	**60.83**	*	*
	[*2*, *5*]		0	251.33	*	*
$\mathbf{Q}_3 =$	*3*			0	*	*
	4				0	*
	5					0

		[*1*, *4*, *3*]	[*2*, *5*]	*3*	*4*	*5*
	[*1*, *4*, *3*]	0	**790.67**	*	*	*
	[*2*, *5*]		0	*	*	*
$\mathbf{Q}_4 =$	3			0	*	*
	4				0	*
	5					0

longer exists as a separate entity. In the jth cell of the first row and column (now the row and column for cluster [1, 4]) is entered

$$q(j, [1,4]) = Q[j,1,4] - Q[j] - Q[1,4].$$

These terms are evaluated for every j not equal to 1 or 4, that is, for $j = 2$, 3 and 5. Recall that for any j value

$$Q[j,1,4] = \tfrac{1}{3}\{d^2(j,1) + d^2(j,4) + d^2(1,4)\},$$

$$Q[j] = 0,$$

and

$$Q[1,4] = \tfrac{1}{2}d^2(1,4).$$

Therefore, letting j take the values 2, 3, and 5 in turn, and taking the required distances² from the matrix $\mathbf{D}^2$, it is found that

$$\begin{aligned} q(2, [1,4]) &= Q[1,2,4] - Q[2] - Q[1,4] \\ &= \tfrac{1}{3}\{d^2(1,2) + d^2(1,4) + d^2(2,4)\} - 0 - \tfrac{1}{2}d^2(1,4) \\ &= \tfrac{1}{3}\{881 + 13 + 1108\} - 0 - \tfrac{13}{2} \\ &= 660.83. \end{aligned}$$

Likewise,

$$\begin{aligned} q(3, [1,4]) &= Q[1,3,4] - Q[3] - Q[1,4] \\ &= 67.33 - 0 - 6.5 \\ &= 60.83 \end{aligned}$$

and $q(5, [1,4]) = 468.83$.

It will be seen that the values just computed appear in the first row of $\mathbf{Q}_2$ (they would also appear in the first column, of course, if the whole matrix were shown, but it is unnecessary to print the matrix in full because it is symmetric).

The remaining elements in $\mathbf{Q}_2$ are the same as in $\mathbf{Q}_1$.

The smallest element in $\mathbf{Q}_2$ is 34, the value of $q[2,5]$. Hence [2, 5] is the second cluster to be formed.

By a similar process, we calculate the terms of $\mathbf{Q}_3$. Since point 5 is no longer separate, the elements in the fifth row and column are replaced with asterisks. The second row and column become the row and column for the new cluster [2, 5], so that the two two-member clusters [1, 4] and [2, 5] now occupy first and second positions in the matrix. The increase in within-cluster dispersion that would result if they were united to make a four-member cluster is

$$q([1,4],[2,5]) = Q[1,2,4,5] - Q[1,4] - Q[2,5]$$

$$= \tfrac{1}{4}\{d^2(1,2) + d^2(1,4) + d^2(1,5) + d^2(2,4)$$

$$+ d^2(2,5) + d^2(4,5)\} - \tfrac{1}{2}d^2(1,4) - \tfrac{1}{2}d^2(2,5) = 830.25.$$

The procedure for computing the elements of the $\mathbf{Q}$ matrices should now be clear.

The smallest element in $\mathbf{Q}_3$ is 60.83 in cell ([1, 4], 3). Therefore, the next fusion creates the cluster [1, 4, 3].

The gain in within-cluster dispersion produced by the final fusion between [1, 4, 3] and [2, 5] is 790.67, the only numerical element in $\mathbf{Q}_4$. It is found from the relation

$$q([1,3,4],[2,5]) = Q[1,2,3,4,5] - Q[1,3,4] - Q[2,5].$$

After the final fusion, when all five points have been united into one cluster, the within-cluster dispersion is

$$Q[1,2,3,4,5] = \frac{1}{5}\sum_{j<k} d^2(j,k) = 892,$$

as already derived at the beginning of this section.

To summarize, let us consider the step-by-step increases that took place in the total within-cluster dispersion (hereafter called the total dispersion) as clustering proceeded. At the start, there were five separate points (or one-member clusters), all with zero within-cluster dispersion. Therefore, the total dispersion was zero. Formation of cluster [1, 4] raised the total dispersion by the amount of its own within-cluster dispersion, namely, 6.5. Likewise, formation of cluster [2, 5] added 34 to this total, bringing it to 40.5. Next, formation of cluster [1, 3, 4] brought an increment of 60.83 to the total (recall that the elements in the criterion matrices are the increases in dispersion that the different possible fusions would bring about, not the within-cluster dispersions themselves). The final fusion of [1, 3, 4] and [2, 5] brought an increment of 790.67. Thus in numbers,

$$6.5 + 34 + 60.83 + 790.67 = 892.$$

In words: the total dispersion is the sum of the smallest elements in the successive criterion matrices, the Qs. The clustering strategy consists in augmenting the total by the smallest amount possible at each step.

Figure 2.8 shows the data points of Data Matrix #3 and the dendrogram we have just computed. The height of each node is the within-cluster

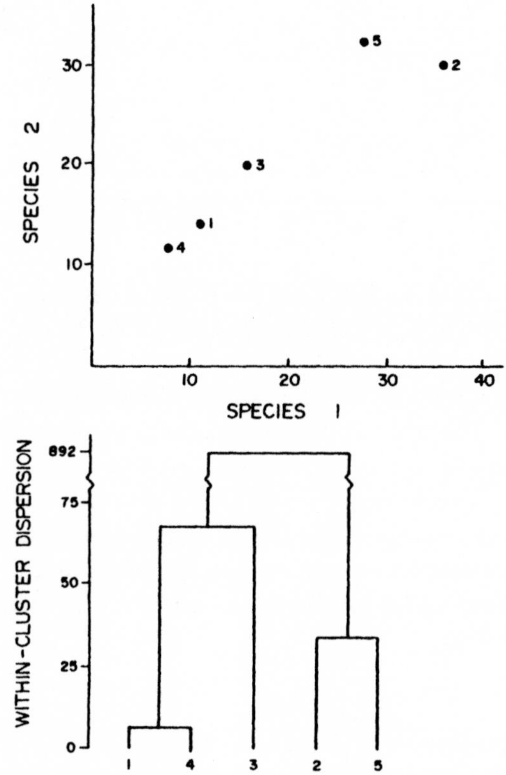

Figure 2.8. The data points of Data Matrix #3 (see Table 2.8) and the dendrogram yielded by minimum variance clustering of the data.

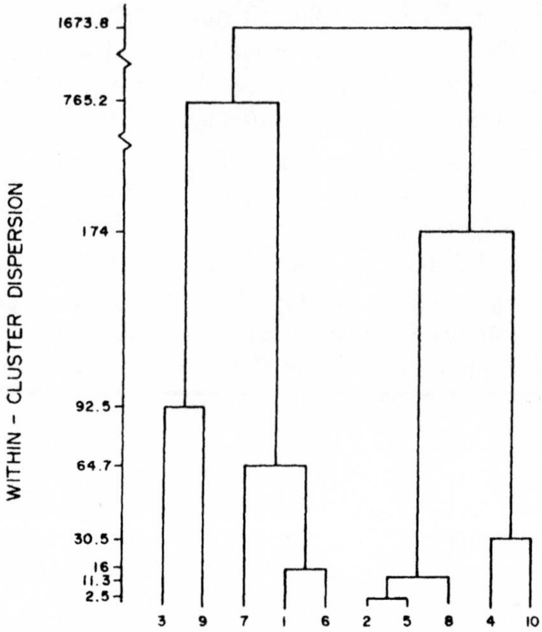

Figure 2.9. The dendrogram produced by applying minimum variance clustering to Data Matrix #1. The scale on the left shows the within-cluster dispersion of each node.

dispersion of the newly formed cluster that the node represents. Thus the heights are $Q[1,4] = 6.5$, $Q[2,5] = 34$, $Q[1,4,3] = 67.33$, and $Q[1,2,3,4,5] = 892$.

Figure 2.9 shows the results of applying minimum variance clustering to Data Matrix #1. The steps in the computations are not shown here since nine 10×10 matrices would be required. The exact value of the within-cluster dispersion of each newly formed cluster is shown on the scale to the left of the dendrogram to serve as a check for readers who wish to carry out minimum variance clustering on these data for themselves. It is interesting to compare this dendrogram with those in Figures 2.3, 2.5, and 2.6 which all relate to the same data.

2.6. DISSIMILARITY MEASURES AND DISTANCES

It was remarked in Section 1 of this chapter that the Euclidean distance between the points representing two quadrats is only one of many possible ways of defining the dissimilarity of the two quadrats. We used Euclidean distance as a dissimilarity measure in Sections 2, 3, 4, and 5 in which four different clustering procedures were described. This section describes some other possible dissimilarity measures and their advantages and disadvantages.

Metric and Nonmetric Measures

First it must be noticed that dissimilarity measures are of two kinds, *metric* and *nonmetric*; the distinction between them is very important.

A metric measure or, more briefly, a metric has the geometric properties of a distance. In particular, it is subject to the *triangle inequality axiom*. This is the common-sense axiom which states that the length of any one side of a triangle must be less than the sum of the lengths of the other two sides. Suppose we write $d(A, B)$ for the length of side AB of triangle ABC, and analogously for the other two sides. Then the triangle inequality may be written

$$d(A, B) \leq d(A, C) + d(B, C).$$

The equality sign applies when A, B, and C are in a straight line or, equivalently, when triangle ABC has been completely flattened to a straight line.

The triangle inequality is obviously true of Euclidean distances. However, measures of the dissimilarity of the contents of two quadrats (or sampling units of any appropriate kind) are often devised without any thought of the geometrical representation of the quadrats as points in a many-dimensional coordinate frame. These measures were not, when first invented, thought of as distances. Only subsequent examination shows whether they are metric, that is, whether they obey the triangle inequality or, in other words, "behave" as distances.

Some examples are given after we have considered why metric dissimilarity measures are to be preferred to nonmetric ones. As remarked previously, when a metric measure is used to define the dissimilarity between two quadrats, then the dissimilarities behave like distances. As a result, it may be possible (sometimes) to plot the quadrats as points in a space of many dimensions with the distance between every pair of points being equal to the dissimilarity of the pair (see page 165). But when a nonmetric dissimilarity measure is used, this cannot be done.

Of course, if Euclidean distance as already defined were used as the dissimilarity measure, then the pattern of points would be the same as that produced when each point has as its coordinates the amounts of the different species in the quadrat it represents. But if some other metric dissimilarity measure were used, it would give a different pattern of points. However, if a nonmetric dissimilarity measure were used, no swarm of points could be constructed of any pattern whatever.

To see this, let us invent a dissimilarity measure simply for purposes of illustration. Suppose we define the dissimilarity between points A and B as

$$\delta(A, B) = \frac{100}{\max(d) - d(A, B)}.$$

Here $d(A, B)$ is the ordinary Euclidean distance as previously used in this chapter, and $\max(d)$ is the distance separating the farthest pair of points. For concreteness, let $\max(d) = 100$. Obviously, increasing values of $d(A, B)$ within the observed range of 0 to 100 give increasing values of $\delta(A, B)$ and, therefore, $\delta(A, B)$ could reasonably be used as a measure of dissimilarity.

Now imagine three points, A, B, and C. Let the Euclidean distance between each pair be

$$d(A, B) = 90; \qquad d(A,C) = 75; \qquad d(B,C) = 50.$$

These distances conform with the triangle inequality; that is,

$$d(A, B) \leq d(A,C) + d(B,C)$$

and, therefore, the points can be plotted in a two-dimensional space (for instance, a sheet of paper) in the form of a triangle with sides 90, 75, and 50. Now consider the dissimilarities defined previously:

$$\delta(A, B) = \frac{100}{100 - d(A, B)} = \frac{100}{100 - 90} = 10;$$

Likewise, $\delta(A,C) = 4$ and $\delta(B,C) = 2$. Clearly, it is *not* true that

$$\delta(A, B) \leq \delta(A,C) + \delta(B,C)$$

and, as a consequence, one can*not* construct a triangle with $\delta(A, B)$,

$\delta(A,C)$, and $\delta(B,C)$ as its sides. It is impossible. This is another way of saying that the δs, although they could serve as dissimilarity measures, are nonmetric.

To repeat, the merit of metric dissimilarity measures is that they often permit the quadrats to be represented as a swarm of points in many-dimensional space. Such a representation is not strictly necessary if all we want to do with the data is classify them. Nearest-neighbor and farthest-neighbor clustering, as examples, can be done just as well with nonmetric dissimilarities as with metric. But often, indeed usually, we want to ordinate the data as well as classify them. As we see in Chapter 4, ordination procedures use swarms of data points as their raw material. Obviously, it is desirable that the two methods of analysis, ordination and classification (or clustering), be carried out on identical bodies of data, that is, on identical swarms. Hence metric dissimilarity measures are to be preferred to nonmetric ones. Their use permits a clustering procedure and an ordination to be performed on the same swarm of data points.

The following are examples of two dissimilarity measures which do not, at first glance, look very different; however, one is metric and the other nonmetric.

The better known of the two is the nonmetric measure *percentage dissimilarity* PD (also known as the *percentage difference* or *percentage distance*). It is the complement of *percentage similarity* PS (also known as *Czekanowski's index of similarity*; see Goodall, 1978a). Since PS, now to be defined, is a percentage, PD is set equal to 100-PS.

The percentage similarity of a pair of quadrats, say quadrat 1 and quadrat 2, is defined as follows.

Let the number of species found in one or both quadrats be s.

Let x_{i1} and x_{i2} be the amount of species i in quadrats 1 and 2, respectively ($i = 1, 2, \ldots, s$). Then

$$\text{PS} = 200 \times \frac{\sum_{i=1}^{s} \min(x_{i1}, x_{i2})}{\sum_{i=1}^{s} (x_{i1} + x_{i2})}. \tag{2.8}$$

A numerical example is shown in Table 2.10. Since, as shown in the table, PS = 58.67%, the percentage dissimilarity is PD = 41.33%.

TABLE 2.10. TO ILLUSTRATE CALCULATION OF THE PERCENTAGE DISSIMILARITY PD AND THE PERCENTAGE REMOTENESS PR OF TWO QUADRATS.[a]

Species Number i	Quadrat 1	Quadrat 2	$\min(x_{i1}, x_{i2})$	$\max(x_{i1}, x_{i2})$
1	25	7	7	25
2	40	16	16	40
3	18	50	18	50
4	16	22	16	22
5	9	22	9	22
Totals	108	117	66	159

PS = 200 × 66/(108 + 117) = 58.67%. Therefore, PD = 41.33%.
RI = 100 × 66/159 = 41.51%. Therefore, PR = 58.49%.

[a] The entries in the table are the quantities of each species in each quadrat.

Calculation of the second dissimilarity measure mentioned, the metric measure, is also shown in Table 2.10. There appears to be no accepted name for it, so it is here called *percentage remoteness*, PR. It is the complement of *Ružička's index of similarity*, RI (Goodall, 1978a), which is

$$\mathrm{RI} = 100 \times \frac{\sum_{i=1}^{s} \min(x_{i1}, x_{i2})}{\sum_{i=1}^{s} \max(x_{i1}, x_{i2})}. \tag{2.9}$$

Then

$$\mathrm{PR} = 100 - \mathrm{RI}.$$

Both PD and PR take values in the range 0 to 100. It is easily seen that if the two quadrats have no species in common, then all terms of the form $\min(x_{i1}, x_{i2})$ are zero and thus PD = PR = 100%. At the other extreme, if the contents of the two quadrats are identical so that $x_{i1} = x_{i2}$ for all i, then PD = PR = 0%. Therefore, either measure could be used as a measure of dissimilarity and if it were not for the superiority of metric over nonmetric measures, there would be little to choose between them. However, since PR is metric, it is superior. A proof that PR is metric can be found in Levandowsky and Winter (1971), and a demonstration that PD is nonmetric can be found in Orlóci (1978) (but see page 57 of this book). An example of the use of PR in ecological work has been given by Levandowsky (1972); he used it to measure the dissimilarity of the phytoplankton in water samples collected from temporary beach ponds on the shores of Long Island Sound.

Another metric dissimilarity measure that has much to commend it is the *city-block distance* CD (sometimes called the *Manhattan metric*). It is the sum of the differences in species amounts, for all species, in the two sampling units being compared. In symbols,

$$\text{CD} = \sum_{i=1}^{s} |x_{i1} - x_{i2}|;$$

here $|x_{i1} - x_{i2}|$ denotes the absolute magnitude of the difference between x_{i1} and x_{i2} taken as positive irrespective of the sign of $(x_{i1} - x_{i2})$. Thus for the data in Table 2.10,

$$\begin{aligned}\text{CD} &= |25 - 7| + |40 - 16| + |18 - 50| + |16 - 22| + |9 - 22| \\ &= 18 + 24 + 32 + 6 + 13 = 93.\end{aligned}$$

In a way, CD is the most intuitively attractive of the dissimilarity measures. It amounts to a numerical value for the difference an observer consciously sees on looking at two sampling units, for example, two quadrats in a salt marsh, two trays full of benthos from Surber samplers, or two plots in a forest. Thus suppose one were to inspect two forest plots and count the number of trees of each species in each plot; then, for many people, the spontaneous answer to the question, "How great is the difference between the plots?" might well be arrived at simply by adding together the differences between the plots in species content, taking all species into account. This is the city block distance and it has the metric property.

The units in which city-block distance and Euclidean distance are measured are the same as the units in which species quantities are measured and, as explained before (page 11), vary from one type of community to another. When an author gives a numerical value for a dissimilarity in a research paper, the units are almost always omitted. Purists may disapprove, but the custom seems to be universal, and leads to no misunderstanding provided all units are fully and clearly defined at the outset. With percentage measures such as PD and PR the problem of units does not arise.

We have now considered three metric measures of dissimilarity: Euclidean distance, percentage remoteness, and city-block distance. Often it is convenient to use the square of Euclidean distance rather than the distance itself as the clustering criterion; that is, at each step of a clustering process, the two clusters are united for which the square of the distance separating them is least. This was done in the example in Section 2.6.

It should be noticed that although it is legitimate to use distance^2 as a clustering criterion, this is not equivalent to using distance^2 as a dissimilarity measure since distance^2 is nonmetric. To see this, consider a numerical example. It is easy to construct a triangle with sides 3, 4, and 6 units long, since $6 < 3 + 4$. But, obviously, one cannot construct a triangle with sides 3^2, 4^2, and 6^2 units long since $6^2 > 3^2 + 4^2$.

Euclidean distance, city-block distance, and percentage remoteness provide a more than adequate armory of dissimilarity measures for use whenever nonnormalized distances are required. It is now necessary to consider the topic of *normalized* versus nonnormalized ("raw") data and to consider whether, and if so how, ecological data and dissimilarity measures derived from them should be normalized for analysis.

Raw versus Normalized Data

Data are said to be normalized when every point is placed at the same distance from the origin of the coordinates so that all that distinguishes the points from one another is their direction from the origin. This is equivalent to disregarding the absolute quantities of each species and considering only the relative quantities.

To see why this is sometimes thought desirable, consider Figure 2.10. The points A, B, and C represent three quadrats laid down in a two-species community and it is clear that $d(A, B) < d(B, C)$. However, the relative proportions of the two species in quadrats B and C are identical. Their great dissimilarity, as measured by $d(B, C)$, arises solely from the fact that the total quantity of the two species combined is much greater in C than in B. Conversely, the very slight dissimilarity represented by the short distance

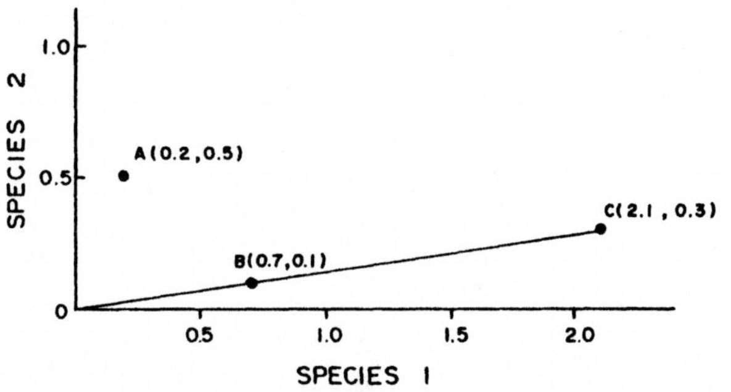

Figure 2.10. Points A, B, and C described in the text. Clearly, $d(A, B) < d(B, C)$.

$d(A, B)$ arises from the fact that both the quadrats A and B contain small amounts of both species; although the ratio of species 1 to species 2 is much greater in B than in A, this difference is not reflected in their dissimilarity as measured by $d(A, B)$.

It can be argued that dissimilarity should be measured in a way that lays more stress on the relative proportions of the species in a quadrat and correspondingly less stress on absolute quantities, in other words, that the raw observations should be normalized. However, this is a matter of opinion and is one of the decisions that must be made before data are analyzed. It should be emphasized that there is no single answer to the question: Should data be normalized before analysis? Whatever the decision, it is a subjective choice. Some guidance towards making the choice is offered in the following. First we consider two ways of measuring the dissimilarity between a pair of data points so that only the relative proportions, not the absolute amounts, of the species are taken into account.

These dissimilarities are the *chord distance* and the *geodesic metric*. Both are metric measures. They are shown diagramatically in Figure 2.11. As always, the simple, two-dimensional (two species) case is used for illustration, and the resulting formulas are then generalized to the s-dimensional (s species) case.

The *chord distance* is derived as follows (see Figure 2.11). Let the original data points be projected onto a circle of unit radius and write A' and B' for the projections of points A and B. Then $d(A', B')$, the Euclidean distance between A' and B', is the chord distance between A and B, which we shall denote by $c(A, B)$. In the figure, since point C represents a quadrat in which the species are present in the same relative proportions as in quadrat

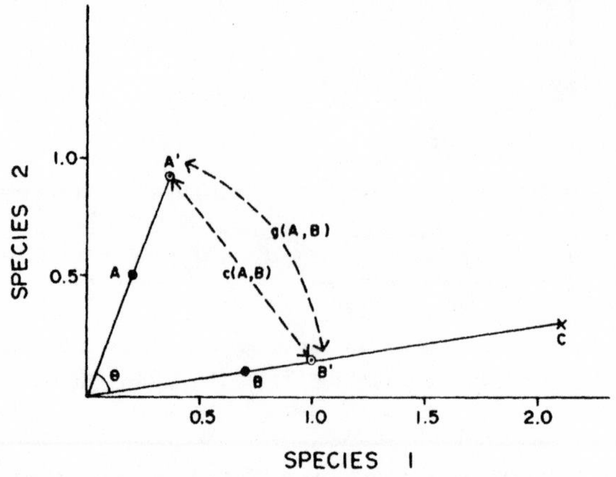

Figure 2.11. Points A, B, and C are the same as in Figure 2.10. A' and B' are the projections of A and B onto a circle of unit radius. The chord distance and the "geodesic metric" (or geodesic distance) separating A and B are $c(A, B)$ and $g(A, B)$, respectively.

B, the point C' is identical with B' so that $d(A',C') = d(A', B')$ and $d(B',C') = 0$. Equivalently, $c(A,C) = c(A, B)$ and $c(B,C) = 0$.

We now derive $c(A, B)$ in terms of the coordinates of points A and B. Let these coordinates be (x_{1A}, x_{2A}) for point A and (x_{1B}, x_{2B}) for point B. (Recall that the first subscript always refers to the species and the second to the quadrat or other sampling unit.)

First, for brevity, put $OA = a$, $OA' = a'$, $OB = b$, and $OB' = b'$. Write θ for angle $\widehat{AOB}$. Obviously, $\widehat{AOB} = \widehat{A'OB'}$.

From the construction, we know that $a' = b' = 1$.

Applying Apollonius's theorem (see page 78) to ΔABO and $\Delta A'B'O$, respectively, shows that

$$d^2(A, B) = a^2 + b^2 - 2ab\cos\theta \tag{2.10}$$

and

$$\begin{aligned} c^2(A, B) &= a'^2 + b'^2 - 2a'b'\cos\theta \\ &= 1^2 + 1^2 - 2\cos\theta \\ &= 2(1 - \cos\theta). \end{aligned} \tag{2.11}$$

Now use (2.10) to express $\cos\theta$ in terms of x_{1A}, x_{2A}, x_{1B}, and x_{2B}.

From Pythagoras's theorem,

$$a^2 = x_{1A}^2 + x_{2A}^2; \qquad b^2 = x_{1B}^2 + x_{2B}^2;$$

$$d^2(A, B) = (x_{1A} - x_{1B})^2 + (x_{2A} - x_{2B})^2.$$

Then since from (2.10),

$$\cos\theta = \frac{a^2 + b^2 - d^2(A, B)}{2ab},$$

it follows that

$$\cos\theta = \frac{(x_{1A}^2 + x_{2A}^2) + (x_{1B}^2 + x_{2B}^2) - [(x_{1A} - x_{1B})^2 + (x_{2A} - x_{2B})^2]}{2\sqrt{(x_{1A}^2 + x_{2A}^2)(x_{1B}^2 + x_{2B}^2)}}$$

$$= \frac{x_{1A}x_{1B} + x_{2A}x_{2B}}{\sqrt{(x_{1A}^2 + x_{2A}^2)(x_{1B}^2 + x_{2B}^2)}}. \tag{2.12}$$

Hence to evaluate $c^2(A, B)$ for any pair of points A and B with known coordinates, first find $\cos\theta$ using (2.12) and then substitute the result in (2.11).

For example, in Figure 2.11 the coordinates of A and B are, respectively,

$$(x_{1A}, x_{2A}) = (0.2, 0.5) \quad \text{and} \quad (x_{1B}, x_{2B}) = (0.7, 0.1).$$

Therefore,

$$\cos\theta = \frac{(0.2 \times 0.7) + (0.5 \times 0.1)}{\sqrt{(0.2^2 + 0.5^2)(0.7^2 + 0.1^2)}} = 0.4990,$$

from (2.12), whence $c^2(A, B) = 1.0021$ and $c(A, B) = 1.0010$ from (2.11); also, $\theta = 1.0484$ radians or 60.07°.

The equations can be directly generalized to the s species case when the data points form an unvisualizable swarm in a conceptual space of s dimensions (a coordinate frame with s mutually perpendicular axes). Thus in the s species case (2.12) becomes

$$\cos\theta = \frac{x_{1A}x_{1B} + x_{2A}x_{2B} + \cdots + x_{sA}x_{sB}}{\sqrt{\left(x_{1A}^2 + x_{2A}^2 + \cdots + x_{sA}^2\right)\left(x_{1B}^2 + x_{2B}^2 + \cdots + x_{sB}^2\right)}}$$

$$= \frac{\sum_{i=1}^{s} x_{iA}x_{iB}}{\left\{\sum_{i=1}^{s} x_{iA}^2 \sum_{i=1}^{s} x_{iB}^2\right\}^{1/2}}.$$

Equation (2.11) is as already given, whatever the number of species.

The maximum and minimum possible values for the chord distance between a pair of points in a space of any number of dimensions are $\sqrt{2}$ and 0, respectively. This follows from (2.11) and the fact that $\cos\theta$ must lie in the range $[-1, 1]$. Thus when OA and OB are parallel, $\cos\theta = 1$, $c^2(A, B) = 0$, and $c(A, B) = 0$; when OA and OB are perpendicular to each other, $\cos\theta = 0$, $c^2(A, B) = 2$, and $c(A, B) = \sqrt{2}$.

Another obvious dissimilarity measure is the *geodesic metric*, shown as $g(A, B)$ in Figure 2.11. It is the distance from A' to B' along the arc of the unit circle; to be exact, one should stipulate that the distance is to be measured along the shorter arc, not the longer arc formed by the rest of the circle. It is seen that, since the circle has unit radius, the arc distance $g(A, B)$ is the same as angle θ measured in radians. To find the angle, one must first evaluate $\cos\theta = S_{AB}$, say, and then find $g(A, B)$ from $g(A, B) = \arccos S_{AB}$.

S_{AB} is known as the cosine separation of the quadrats (Orlóci, 1978; p. 199). In the two-species example considered previously and shown in Figure 2.11, where the coordinates of A and B are (0.2, 0.5) and (0.7, 0.1), respectively, $\cos\theta = S_{AB} = 0.4990$, as already determined. Hence,

$$g(A, B) = \arccos S_{AB} = 1.0484 \text{ units of length.}$$

The range of possible values of θ, and hence of $g(A, B)$, is from 0 to $\pi/2 = 1.576$. In the simple two-species case illustrated in Figure 2.11 the geodesic metric is the length of an arc of the unit circle. In the three species (three-dimensional) case the metric is the length of a geodesic on a sphere of unit radius and the word *geodesic* has its customary meaning, namely, the

TABLE 2.11. DATA MATRIX #4. THE QUANTITIES OF TWO SPECIES IN ELEVEN QUADRATS.

Quadrat	1	2	3	4	5	6	7	8	9	10	11
Species 1	3	4	5	5.5	6	6	11	11.5	12	14	13.5
Species 2	3	7	7	5.5	4	6.5	11	13.5	13	11	15

shortest on-the-surface distance, or great circle distance, between two points on a sphere. In the s species case the geodesic metric is a great circle on an s-dimensional hypersphere.

Example. We now examine the outcomes of clustering the same set of data twice using the centroid clustering method both times. First, the data are left in raw form and Euclidean distance is used as the dissimilarity measure. Second, the data are normalized and the geodesic metric is used as the dissimilarity measure. The data (Data Matrix #4) are tabulated in Table 2.11 and plotted in Figure 2.12.

There are two "natural" clusters but they differ from each other chiefly in the quantities of the two species they contain; the relative proportions of the species are not very different. Thus if the points represented randomly placed vegetation quadrats, one would infer that the area sampled was a mosaic of fertile areas and sterile areas, but that the vegetation in these two areas differed mainly in its abundance and hardly at all in its species composition. As one would expect, if clustering is done using Euclidean distance as the dissimilarity measure (upper dendrogram in the figure), the two natural clusters are separated clearly; whereas if one uses the geodesic metric as the dissimilarity measure (lower dendrogram), the two clusters are intermingled.

Which is "better" is obviously a matter of choice. Even the meaning of "better" is undefined unless the investigator has some definite object in view, that is, some clearly formulated question for which an answer is sought. Then whatever clustering method gives an unambiguous answer to the question (if any method does) is obviously the best.

Communities which differ from place to place only in overall abundance, and not at all in species composition, are most unlikely to be found in nature. For instance, the abundance of the marine macroalgae on a rocky seashore varies markedly with the exposure of the shore to waves; sheltered shores support a much larger crop than wave-battered shores. But these contrasted communities are not sparse and dense versions of the same species mixture; they differ, also, in species composition.

Likewise, the luxuriance of the ground vegetation in regions severely affected by air pollution is conspicuously less than that in clean areas. But it is not only less in amount; it is also much poorer in species.

Thus if clustering is done to disclose differences of an unspecified kind, raw data are better than normalized data. Differences in overall abundance are not "meaningless" and are not (usually) unaccompanied by at least some qualitative differences in the community of interest. Normalizing the data may inadvertently obscure real, but slight, differences among them at the same time as it (intentionally) obliterates the quantitative differences.

That is not to say, however, that there may not be situations in which normalization is called for; for example, one might wish to classify sample

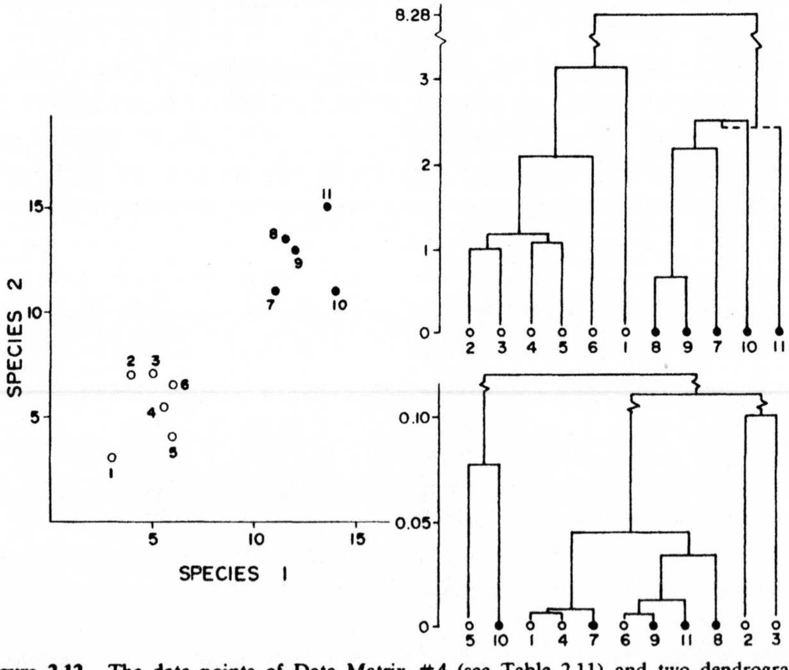

Figure 2.12. The data points of Data Matrix #4 (see Table 2.11) and two dendrograms obtained by clustering the data. Centroid clustering was used for both. The upper dendrogram was obtained using the Euclidean distance between each pair of raw data points as measure of between-quadrat similarity; the lower dendrogram used the geodesic metric.

plots of the vegetation of a polluted area so as to disclose the probable prepollution clustering. Then, provided the qualitative differences among preexisting clusters exceeded the qualitative differences induced by pollution, normalization would be desirable. It might prevent differences in the quantity of vegetation in the sample plots from overriding, and masking, the qualitative differences that persisted from the prepollution period.

To repeat: whether to use raw or normalized data is always a matter of judgment.

Presence-and-Absence Data

In some ecological investigations, it often seems better simply to list the species present in each sampling unit than to attempt to measure or estimate the quantities. When this is done, the resulting data are known as presence–absence data, *binary data*, or (0, 1) data, and the elements in the data matrix consist entirely of 0s and 1s.

Suppose the community being sampled appears to vary appreciably from place to place; then for a given outlay of time and effort one may be able to acquire a larger amount of information, or more useful information, by examining many quadrats quickly rather than a few quadrats slowly and carefully; the quickest way to record a quadrat's contents is, of course, just to list the species in it. Again, suppose the organisms comprising the community vary enormously in size. They might range from tall trees to dwarf shrubs, for example. It might then be impossible to find a quadrat size that was large enough for use with the trees and, at the same time, small enough for it to be practicable to measure the amounts of each species of ground vegetation. In such a case, use of binary data overcomes the difficulty. As Goodall (1978a) has written, in highly heterogeneous communities, "quantitative measures add little useful information" to that yielded by a simple species list for each quadrat.

Now consider the graphic representation of a binary data matrix. In the simple, visualizable two and three-species cases, all the data points must fall on the vertices of a square or a cube, respectively. This amounts to saying that there are, respectively, only four or eight possible positions for the data points in these cases. Thus in the two-species case the coordinates of the four possible data points are (0, 0), (0, 1), (1, 0), and (1, 1). In the three-species case, the eight possible data points have coordinates (0, 0, 0), (1, 0, 0), (0, 1, 0), (0, 0, 1), (1, 1, 0), (1, 0, 1), (0, 1, 1), and (1, 1, 1) (see Figure 2.13).

Now extend the argument to binary data from an s-species community plotted (conceptually) in an s-dimensional coordinate frame. It is intuitively clear that the possible data points are the vertices of an s-dimensional hypercube and the number of these vertices is 2^s.

There is no objection to using the clustering methods described in earlier sections of this chapter to the clustering of binary data. However, unless the total number of species is large, the result of a clustering process may seem somewhat arbitrary. This is because only a few values are possible for the distance separating any pair of points.

Consider the simple cases in Figure 2.13. In the two-species case the distance between any pair of noncoincident points must have one of two values, 1 or $\sqrt{2}$, depending on whether the two points are at the ends of a

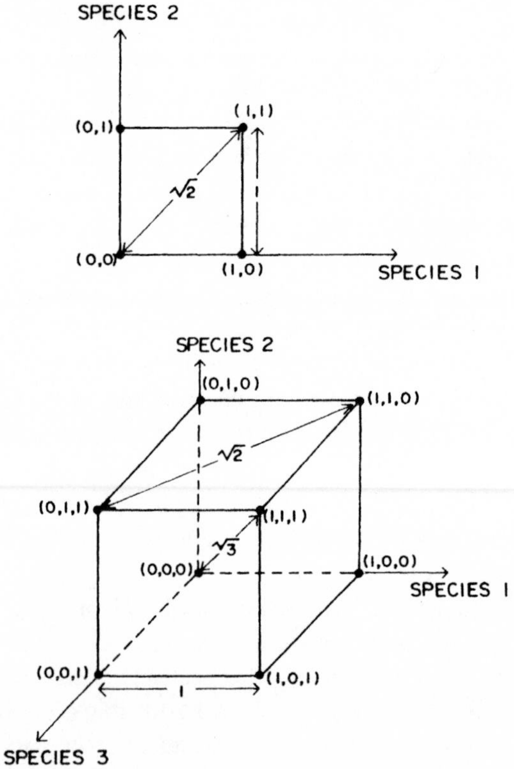

Figure 2.13. All possible binary data points, and the distances separating them, in two-dimensional (upper) and three-dimensional (lower) coordinate frames.

side of the square or at the ends of a diagonal. In the three-species case there are three possible nonzero distances: these are 1, if the two points are at the ends of one edge of the cube; $\sqrt{2}$ if they are at the ends of a diagonal of one of the square faces of the cube; $\sqrt{3}$ if they are at the ends of a diagonal that crosses the cube. Again the argument can be generalized to the s-species case. When there are s species, the distance between any pair of noncoincident points must have one of only s distinct values, namely, $1, \sqrt{2}, \sqrt{3}, \dots, \sqrt{s}$. The distance is the square root of the number of species found in one or the other (but not both) of the quadrats represented by the points. Thus the distance between the points $(1,0,1,0,1,1,1,1,0)$ and $(0,1,0,0,1,1,1,1,1)$ in nine-dimensional space is $\sqrt{4} = 2$, since there are four mismatches between these two lists. In the s-species case, if the two quadrats together contain all s species but have no species in common, then the distance between the points representing them is $\sqrt{s}$.

It follows that if a clustering procedure starts with construction of a distance matrix (e.g., like that in Table 2.1, page 18), whose elements are the distances between every possible pair of points, then unless the number of species is very large there are likely to be several "ties" in the matrix. That is, several elements may all have the same value. If this also happens to be the smallest value, then several fusions become "due" simultaneously. The same thing happens with minimum variance clustering; if two or more elements in a criterion matrix (such as $\mathbf{Q}_1$ in Table 2.9, page 36) are equal to one another and smaller than all the others, then again the indicated fusions are due simultaneously. When this happens, the "due" fusions should be carried out simultaneously before the next distance matrix (or criterion matrix) is constructed; otherwise, errors will occur.

We now consider other ways of measuring the dissimilarity between pairs of quadrats (data points) when the data are in binary or $(0,1)$ form. The Euclidean distance between two points which, as we have seen, is always the distance between two vertices of a hypercube, is not the only way of measuring the dissimilarity of the points. One can also use percentage dissimilarity PD and percentage remoteness PR which were defined earlier (see page 43).

These measures can be calculated as already shown in Table 2.10 (page 44); alternatively, they can be derived from a 2×2 table as we shall now see. Suppose a 2×2 table is constructed to permit two chosen quadrats (quadrats 1 and 2, say) from a sample of several quadrats to be compared. Assume that species lists have been compiled for all the quadrats sampled, and some quadrats contain species that are not present in quadrats 1 and 2. In other words, of the total of s species represented in the data matrix some are absent from both the quadrats being compared. Hence these "joint absences" can be specified and counted. Now consider the following 2×2 table:

		Quadrat 2 Number of species	
		Present	Absent
Quadrat 1 Number of species	Present	a	b
	Absent	c	d

Recall Equation (2.8) on page 43, the definition of percentage similarity PS. Clearly, when x_{i1} and x_{i2} are either 0 or 1 for all values of i (i.e., for all s species),

$$\sum_{i=1}^{s} \min(x_{i1}, x_{i2}) = a, \tag{2.13}$$

the number of species present in both quadrats.

Similarly,

$$\sum_{i=1}^{s} (x_{i1} + x_{i2}) = (a + b) + (a + c) = 2a + b + c; \tag{2.14}$$

this is the number of species in quadrat 1 plus the number in quadrat 2, counting the a "joint presences" (species present in both quadrats) twice over.

Substituting from (2.13) and (2.14) into (2.8) gives

$$\text{PS} = 200 \times \frac{a}{2a + b + c} = 100 \times \frac{2a}{2a + b + c}$$

as the percentage similarity between two quadrats when the data are in binary form. This is identical with *Sørensen's similarity index* (as a percentage), one of the best known and most widely used of the similarity indices available to ecologists. It follows that, with binary data, the percentage dissimilarity PD is the complement of Sørensen's index.

Next recall (2.9), the formula for Ružička's similarity index RI (page 44). The term in the denominator is

$$\sum_{i=1}^{s} \max(x_{i1}, x_{i2}) = a + b + c; \tag{2.15}$$

this is the number of species in the two quadrats combined, *not* counting the joint presences twice.

Substituting from (2.13) and (2.15) into (2.9) gives

$$\text{RI} = 100 \times \frac{a}{a + b + c}.$$

This is *Jaccard's index* (as a percentage), the oldest similarity index used by ecologists (Goodall, 1978a) and as well known as Sørensen's. Thus with binary data the percentage remoteness PR is identical with the complement of Jaccard's index. Indeed, this complement (as a proportion rather than a percentage), namely,

$$1 - \frac{a}{a + b + c} = \frac{b + c}{a + b + c}$$

is known as the *Marczewski–Steinhaus distance* (Orlóci, 1978). It is the ratio of the number of "single" occurrences (species in one but not both of the two quadrats being compared) to the total number of species (those in one or other or both of the quadrats).

The numerical example in Table 2.12 illustrates the relationships among the various measures discussed. To summarize: when the data are binary, percentage dissimilarity PD is identical to the complement of Sørensen's index, and percentage remoteness PR is identical to the Marczewski–Steinhaus distance MS. (It is assumed that the measures are either all in the form of percentages or all in the form of proportions.)

This statement enables one to choose wisely between the competing measures. It has already been mentioned (page 45) that PR is metric and PD nonmetric. It follows that MS, which is no more than a particular form of PR, is metric; a proof, which is rather long, has been given by Levandowsky and Winter (1971). Similarly, the complement of Sørensen's index, which is no more than a particular form of PD, is nonmetric; Orlóci (1978, p. 61) demonstrates the truth of this with an example. Hence MS is the better dissimilarity measure of the two.

TABLE 2.12. ILLUSTRATION OF THE CALCULATION OF DISSIMILARITY MEASURES WITH BINARY DATA.

1. The presences and absences of 12 species in 2 quadrats (compare Table 2.10).

Species Number	Quadrat 1	Quadrat 2	$\min(x_{i1}, x_{i2})$	$\max(x_{i1}, x_{i2})$
1	1	1	1	1
2	1	1	1	1
3	0	0	0	0
4	1	1	1	1
5	0	1	0	1
6	1	1	1	1
7	1	1	1	1
8	1	0	0	1
9	0	1	0	1
10	1	1	1	1
11	0	1	0	1
12	0	0	0	0
Totals	7	9	6	10

PS = 200 × 6/(7 + 9) = 75%. Therefore, PD = 25%.
RI = 100 × 6/10 = 60%. Therefore, PR = 40%.

2. The same data in the form of a 2 × 2 table.

		Quadrat 2: Species present	Quadrat 2: Species absent
Quadrat 1	Species present	$a = 6$	$b = 1$
	Species absent	$c = 3$	$d = 2$

Complement of Sørensen's index = $100(b + c)/(2a + b + c) = 25\%$.
MS distance = $100(b + c)/(a + b + c) = 40\%$.

Since MS and the complement of Sørensen's index can both be expressed as functions of the cell frequencies in the 2×2 table given, it is interesting to enquire whether (assuming binary data) the Euclidean distance between two data points can also be expressed in terms of these frequencies. Recall that the distance between quadrats 1 and 2, $d(1,2)$, is the square root of the number of species that occur in one or the other, but not both, of the quadrats. Hence

$$d(1,2) = \sqrt{b + c}\,.$$

We must now compare the Euclidean distance d with the Marczewski–Steinhaus distance MS in an attempt to decide which (if either) is the better. Since both are metric, some other criterion is needed for judging between them.

The distinctive characteristic of MS is that it takes no account of species that are absent from both the quadrats being compared. This is regarded as a great advantage by ecologists who argue that presences and absences should not be given equal weight, especially for a community made up of sessile organisms. A "presence" conveys the unambiguous information that the species concerned can and does occur in the quadrat concerned, but an "absence" may mean either that the species cannot survive in the quadrat or

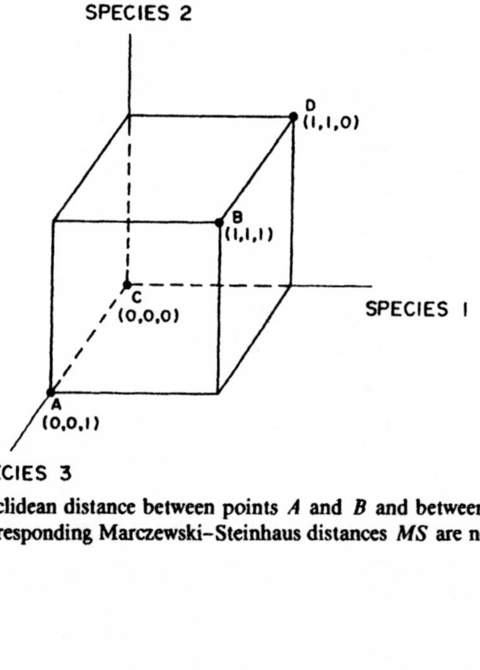

Figure 2.14. The Euclidean distance between points A and B and between points C and D are equal. But the corresponding Marczewski–Steinhaus distances MS are not equal. See text.

that it is absent merely by chance. Thus a dissimilarity measure, or "distance," that ignores joint absences appears to have an advantage.

Euclidean distance treats presences and absences equally, as demonstrated in the following.

The disadvantage of MS, a fatal disadvantage according to Orlóci (1978, p. 62), is that it has no uniform scale of measure. This is most easily seen from Figure 2.14 which shows four data points A, B, C, and D in three-space. Clearly, the distance between points A and B is equal to the distance between points C and D, and both are equal to $\sqrt{2}$. That is,

$$d(A, B) = d(C, D) = \sqrt{2}.$$

This is obvious geometrically. The same result can be derived by constructing 2 × 2 tables for each pair of points; thus

Pair (A, B)

		Quadrat A +	Quadrat A −
Quadrat B	+	1	2
	−	0	0

Pair (C, D)

		Quadrat C +	Quadrat C −
Quadrat D	+	0	2
	−	0	1

(+ denotes presences and − absences).

Hence $d(A, B) = d(C, D) = \sqrt{b + c} = \sqrt{2 + 0} = \sqrt{2}$ as before.

Now consider the MS distances, say, MS(A, B) and MS(C, D), between the two pairs of points. From the frequencies in the 2 × 2 tables,

$$\text{MS}(A, B) = \frac{b + c}{a + b + c} = \frac{2 + 0}{1 + 2 + 0} = \frac{2}{3}$$

and

$$\text{MS}(C, D) = \frac{2 + 0}{0 + 2 + 0} = 1.$$

so that MS$(A, B) \neq$ MS(C, D).

The difference between MS(A, B) and MS(C, D) is due to the term a (the number of species present in both quadrats) in the denominator of MS. In general, even if two pairs of points have the same number of "mismatches" $(b + c)$, the pair with the larger number of species for the combined pair ($a + b + c$) will seem to be the "closer" if MS distances are used as measures of dissimilarity.

In spite of the theoretical contrast between Euclidean distance and MS distance, the difference is probably unimportant in practice; it is unlikely to have much effect on the form of the dendrogram produced by a clustering procedure.

EXAMPLE. For instance, Figure 2.15 shows the results of applying nearest-neighbor clustering to Data Matrix #5 (see Table 2.13). The clustering was performed twice, once with Euclidean distances in the distance matrices (see Section 2.2), giving the dendrogram on the left, and once with MS distances in the distance matrices, giving the dendrogram on the right. As may be seen, they are very similar.

To conclude this section, here is a dendrogram showing the way in which the dissimilarity measures described are related. The measures in boldface are metric, those in italics nonmetric. The arrows lead to dissimilarities usable with binary data from their quantitative "parents."

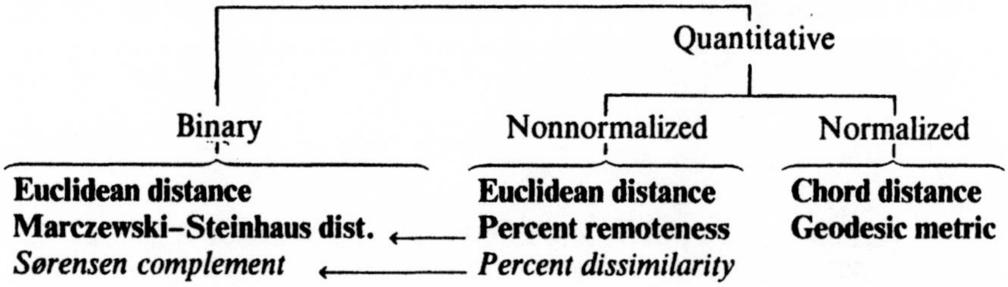

Recall that four of the dissimilarity measures described here are the complements of similarity measures. The way in which they are paired is listed below:

Similarity Measure	Dissimilarity Measure
Percentage similarity, PS	Percentage dissimilarity, PD
Ružička's Index, RI	Percentage remoteness, PR
Jaccard's Index	Marczewski–Steinhaus distance, MS
Sørensen's Index	Complement of Sørensen's Index (no other name has been devised)

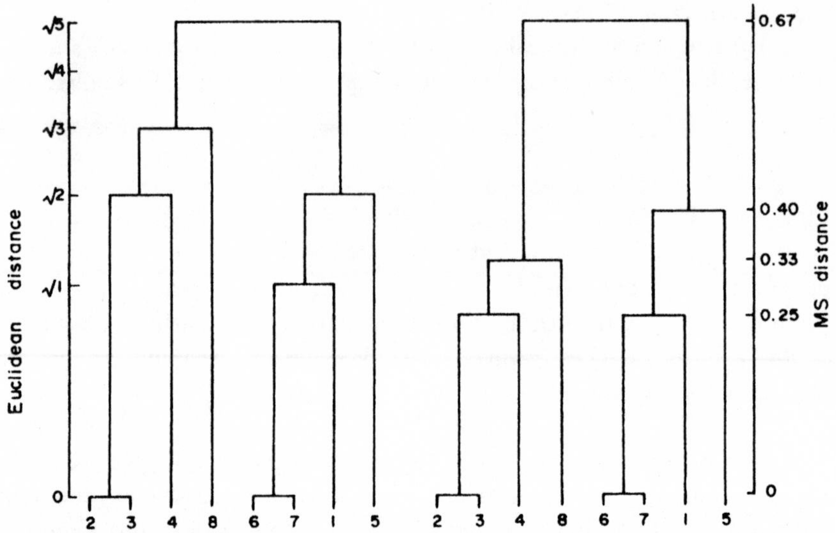

Figure 2.15. Two dendrograms produced by applying nearest-neighbor clustering to Data Matrix #5 (see Table 2.13). Euclidean distance was used as dissimilarity measure for the dendrogram on the left, MS distance for the dendrogram on the right.

TABLE 2.13. DATA MATRIX #5. PRESENCES (1) AND ABSENCES (0) OF 10 SPECIES IN 8 QUADRATS.

Quadrat	1	2	3	4	5	6	7	8
Species 1	0	1	1	1	0	0	0	1
2	0	1	1	0	1	0	0	1
3	0	1	1	1	0	0	0	1
4	1	0	0	0	1	1	1	0
5	1	1	1	1	1	1	1	0
6	1	1	1	1	0	1	1	0
7	1	1	1	0	1	0	0	1
8	0	1	1	1	0	0	0	1
9	0	1	1	1	0	0	0	1
10	0	0	0	0	0	0	0	1

2.7. AVERAGE LINKAGE CLUSTERING

In Section 2.4 it was remarked that there are many possible ways of defining the distance between two clusters. In the clustering method described in that section (centroid clustering) the distance between two clusters is defined as the distance between their centroids. In this section we consider other definitions and their properties.

The Average Distance between Clusters

The most widely used intercluster distance is the *average distance*. This distance is most easily explained with the aid of a diagram; see Figure 2.16.

The five data points with their coordinates given beside them show the amounts of species 1 and 2 in each of five quadrats. There are two obvious clusters, $[A, B, C]$ and $[D, E]$. The average distance between these clusters, which will be written $d_u([A, B, C], [D, E])$, is defined as the arithmetic average of all distances between a point in one cluster and a point in the other. There are six such distances. Therefore,

$$d_u([A,B,C],[D,E]) = \tfrac{1}{6}\{d(A,D) + d(A,E) + d(B,D) + d(B,E) + d(C,D) + d(C,E)\}.$$

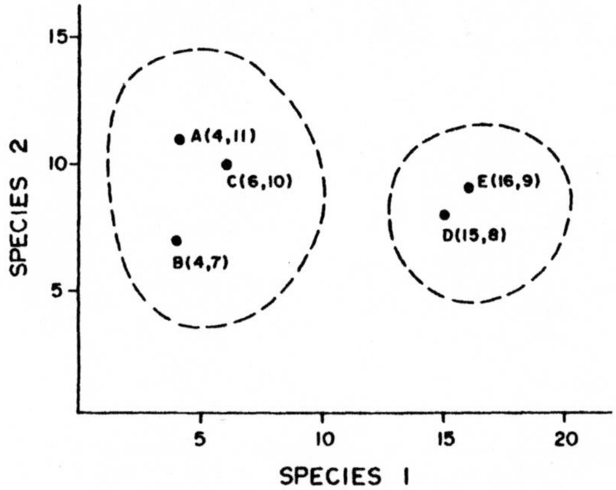

Figure 2.16. Data points showing the quantities of two species in five quadrats, to illustrate the definition of $d_u([A, B, C], [D, E])$. See text.

Here, as always, $d(A, D)$, for example, denotes the Euclidean distance between the two individual points A and D.

Now

$$d(A,D) = \sqrt{(4-15)^2+(11-8)^2} = 11.4018;$$

$$d(A,E) = \sqrt{(4-16)^2+(11-9)^2} = 12.1655;$$

$$\cdots\cdots\cdots\cdots\cdots\cdots\cdots\cdots$$

and

$$d(C,E) = \sqrt{(6-16)^2+(10-9)^2} = 10.0499.$$

It is easily found, after calculating all six interpoint distances, that

$$d_u([A,B,C],[D,E]) = 11.0079.$$

Now consider the general case. We require an equation for the average distance between an m-member cluster $[M_1, M_2, \ldots, M_m]$ and an n-member cluster $[N_1, N_2, \ldots, N_n]$. There are clearly mn point-to-point distances to be averaged. Therefore,

$$d_u([M_1, M_2, \ldots, M_m], [N_1, N_2, \ldots, N_n])$$

$$= \frac{1}{mn}\{d(M_1, N_1) + d(M_1, N_2) + \cdots + d(M_m, N_n)\}$$

$$= \frac{1}{mn}\sum_{j=1}^{m}\left[\sum_{k=1}^{n} d(M_j, N_k)\right]$$

or, equivalently,

$$\frac{1}{mn}\sum_{k=1}^{n}\left[\sum_{j=1}^{m} d(M_j, N_k)\right].$$

Notice that the order of summation is immaterial and the large brackets are unnecessary. Thus we may write

$$d_u([M_1, M_2, \ldots, M_m], [N_1, N_2, \ldots, N_n]) = \frac{1}{mn} \sum_j \sum_k d(M_j, N_k), \tag{2.16}$$

it being understood that the summations are over all values of j and k.

Equation (2.16) is the symbolic form of the definition of average distance. But when these distances are used to decide which pair of clusters should be united at each of the successive steps of a clustering process, it is much more economical computationally to derive each successive intercluster distance-matrix from its predecessor rather than by using (2.16) which expresses each distance in terms of the coordinates of the original data points. This may be done as follows (Lance and Williams, 1966):

Consider three clusters $[M_1, M_2, \ldots, M_m]$, $[N_1, N_2, \ldots, N_n]$, and $[P_1, P_2, \ldots, P_p]$ with m, n, and p members, respectively. In what follows, the clusters are represented by the more compact symbols $[M]$, $[N]$, and $[P]$. Suppose $[M]$ and $[N]$ are united to form the new cluster $[Q]$, with $q = m + n$ members. Then, from (2.16),

$$d_u([P],[Q]) = \frac{1}{pq} \sum_{j=1}^{p} \sum_{k=1}^{q} d(P_j, Q_k) \tag{2.17}$$

Now recall that the points belonging to the new cluster $[Q]$ are $M_1, M_2, \ldots, M_m, N_1, N_2, \ldots, N_n$. Therefore, we can separate the right side of (2.17) into two components and write

$$d_u([P],[Q]) = \frac{1}{pq} \sum_{j=1}^{m} \sum_{k=1}^{p} d(M_j, P_k) + \frac{1}{pq} \sum_{j=1}^{n} \sum_{k=1}^{p} d(N_j, P_k)$$

Now multiply the first term on the right side by $m/m = 1$ and the second term by $n/n = 1$. This maneuver obviously does not alter the value of the expression. That is,

$$d_u([P],[Q]) = \frac{m}{m}\frac{1}{pq}\sum_{j=1}^{m}\sum_{k=1}^{p} d(M_j, P_k) + \frac{n}{n}\frac{1}{pq}\sum_{j=1}^{n}\sum_{k=1}^{p} d(N_j, P_k)$$

$$= \frac{m}{q}\frac{1}{mp}\sum_{j=1}^{m}\sum_{k=1}^{p} d(M_j, P_k) + \frac{n}{q}\frac{1}{np}\sum_{j=1}^{n}\sum_{k=1}^{p} d(N_j, P_k).$$

From (2.16) it is seen that

$$\frac{1}{mp}\sum_{j=1}^{m}\sum_{k=1}^{p} d(M_j, P_k) = d_u([M],[P])$$

and

$$\frac{1}{np}\sum_{j=1}^{n}\sum_{k=1}^{p} d(N_j, P_k) = d_u([N],[P]).$$

Recall that $q = m + n$ and that $[Q]$ contains all the members of $[M]$ and $[N]$, the two clusters that were combined to form it. Thus

$$d_u([P],[Q]) = \frac{m}{m+n} d_u([M],[P]) + \frac{n}{m+n} d_u([N],[P]). \tag{2.18}$$

As a numerical example, consider the points in Figure 2.16 again. When these points are clustered by any method, it is obvious that points D and E will be united first and then points A and C. After these two fusions have been done there are three clusters, which will be labeled as follows:

$[M]$ is the one-member cluster consisting of point B;

$[N]$ is the two-member cluster consisting of points A and C;

$[Q]$ is the three-member cluster consisting of points B, A, and C formed by uniting clusters $[M]$ and $[N]$.

$[P]$ is the two-member cluster consisting of points D and E.

Thus $m = 1$, $n = 2$, $q = 3$, and $p = 2$. From the definition of (2.16),

$$d_u([M],[P]) = \tfrac{1}{2}\{d(B,D) + d(B,E)\} = 11.6054;$$

$$d_u([N],[P]) = \tfrac{1}{4}\{d(A,D) + d(A,E) + d(C,D) + d(C,E)\}$$
$$= 10.7092.$$

Hence from (2.18),

$$d_u([P],[Q]) = \tfrac{1}{3} \times 11.6054 + \tfrac{2}{3} \times 10.7092 = 11.0079.$$

This is, as it should be, the same as $d_u([A, B, C], [D, E])$ as given on page 64.

Unweighted and Weighted Distances As Clustering Criteria

The average distance between clusters $[P]$ and $[Q]$, previously denoted by $d_u([P],[Q])$, is not the only way of measuring intercluster distance. Recall and compare equations (2.18) and (2.7) (page 32). They constitute two different answers to the question: *What is the distance between two clusters given that the first has just been created by the fusion of two preexisting clusters each of which was at a known distance from the second cluster?* (Observe that the question asked is not the simpler one: What is the distance between two clusters? The reason is that the answer sought is a formula for computing the elements of each distance matrix from its predecessor.)

Let us write $[Q]$ for the first cluster, $[M]$ and $[N]$ for the preexisting clusters from which $[Q]$ was formed, and $[P]$ for the second cluster. The numbers of points in these clusters are q, m, n, and p, respectively, with $q = m + n$.

The answer to the preceding question depends on how intercluster distance is defined. As we shall see, the defining equations are sometimes expressed in terms of distance d, and sometimes of distance squared d^2. To make the relationship among the definitions more apparent, the word "dissimilarity" is used here to mean either distance or distance², according to context. The symbol δ is used in the equations to denote either d or d^2, and after each equation its current meaning is specified.

If dissimilarity is defined as the average distance, the answer to the question is given by (2.18), rewritten with δ in place of d, namely,

$$\delta_u([P],[Q]) = \frac{m}{m+n}\delta_u([M],[P]) + \frac{n}{m+n}\delta_u([N],[P]). \quad (2.19)$$

Here δ denotes d.

On the other hand, if dissimilarity is defined as the distance² between cluster centroids (i.e., as the squared *centroid distance*), the answer to the question becomes

$$\delta_c([P],[Q]) = \frac{m}{m+n}\delta_c([M],[P]) + \frac{n}{m+n}\delta_c([N],[P])$$

$$- \frac{mn}{(m+n)^2}\delta_c([M],[N]). \qquad (2.20)$$

This is Equation (2.7) with $x^2 = \delta_c([P],[Q])$, $a^2 = \delta_c([M],[P])$, $b^2 = \delta_c([N],[P])$, and $c^2 = \delta_c([M],[N])$. In (2.20) δ denotes d^2. The subscripts in δ_u and δ_c stand for "unweighted" and "centroid," respectively; δ_u may be described as the *unweighted average distance*.

Both these dissimilarities are described as unweighted because they attach equal weight to every individual point. Therefore, the weight of a cluster is treated as proportional to the number of points it contains. As a result, the centroid (center of gravity) of a pair of clusters is not at the midpoint between the centroids of the separate clusters but is closer to the cluster with the larger number of members (see Figure 2.7, page 30).

We now consider "weighted dissimilarities." These are defined in a way that attaches equal weight to every cluster, and hence unequal weights to the individual points. Therefore, the definitions are very easily obtained by setting $m = n = 1$ in Equations (2.19) and (2.20). Thus from (2.19) we get

$$\delta_w([P],[Q]) = \tfrac{1}{2}\delta_w([M],[P]) + \tfrac{1}{2}\delta_w([N],[P]). \qquad (2.21)$$

Here δ denotes d and the subscript w stands for "weighted"; d_w is the *weighted average distance*.

Similarly, (2.20) is replaced by

$$\delta_m([P],[Q]) = \tfrac{1}{2}\delta_m([M],[P]) + \tfrac{1}{2}\delta_m([N],[P]) - \tfrac{1}{4}\delta_m([M],[N]).$$

$$(2.22)$$

Here δ denotes d^2 and the subscript m stands for "median"; d_m is the *median distance*, sometimes known as the *weighted centroid distance*.

Equation (2.22) can be obtained directly by considering Figure 2.7. If we assume that, whatever the values of m and n, the centroid of the cluster formed by uniting $[M]$ and $[N]$ lies midway between them at distance $c/2$

from each, then it is clear that

$$x^2 = \tfrac{1}{2}a^2 + \tfrac{1}{2}b^2 - \tfrac{1}{4}c^2$$

from which (2.22) follows in the same way that (2.20) follows from (2.7).*

The Four Versions of Average Linkage Clustering

Four ways of measuring intercluster distance have now been described: the unweighted average distance, the weighted average distance, the centroid distance (unweighted), and its weighted equivalent, the median distance. Each of these differently defined distances can be used as the basis of a clustering process. At every step of such a process, the pair of clusters separated by the smallest distance (using whichever definition of distance has been chosen) is united.

The four clustering methods that use these distances are known, collectively, as *average linkage clustering*. Centroid clustering, described in detail in Section 2.4, is one of the four. The relationships among the four are most clearly shown by arraying them in a 2 × 2 table thus (below the name of each method is given the number of the equation to be used in the computations):

	Intercluster Distance	
	Average of Interpoint Distances	Distance Between Centroids
Unweighted	Unweighted group average method (2.19)	Centroid method (2.20)
Weighted	Weighted group average method (2.21)	Median method (2.22)

The methods were named by Lance and Williams (1966).

*For summary definitions of these four measures of the dissimilarity between two clusters, one of which has been formed by uniting two preexisting clusters, the reader is referred to the Glossary; see under *Average Linkage Clustering Criteria.*

Three points should be noticed before examples are given; they are discussed in the following paragraphs.

1. All four methods have the great computational advantage of being *combinatorial* (Lance and Williams, 1966). That is, once the distances between every pair of points in the original swarm of data points have been computed and entered in a distance-matrix, the coordinates of the points are not needed again. Each succeeding distance-matrix is calculable from its predecessor, using the appropriate equation as indicated in the preceding table.

2. All four methods can quite easily be carried out using either d or d^2 in place of δ in Equations (2.19), (2.20), (2.21), and (2.22). Thus each method can be made to yield two different dendrograms since d and d^2 do not give the same results. However, there seems to be no good reason for using d^2 rather than d for either of the group average methods. For the centroid and median methods, on the other hand, d^2 is preferable to d as clustering criterion. As noted on page 32, Equations (2.7) and, likewise, (2.20) and (2.22) with δ set equal to d^2, have a definite geometrical meaning; this is not so if δ is set equal to d. With the centroid and median methods, therefore, it is best to use values of d^2 as clustering criteria (i.e., to unite the cluster pair for which d^2 is a minimum at each step). But this does *not* amount to using d^2 as a dissimilarity measure. Rather, one uses d as dissimilarity measure and its square as clustering criterion.

3. It is worth noticing that the terms "weighted" and "unweighted" have been used here as Lance and Williams (1966) and Sneath and Sokal (1973) use them. Their usage is unexpected and apt to mislead unless it is remembered that the word "unweighted" applies to the original data points (see Gower, 1967).

EXAMPLE. Figure 2.17 shows the dendrograms obtained by applying the four clustering methods to Data Matrix #6 (see Table 2.14). The clustering criterion was d for the *group average methods* and d^2 for the centroid and *median methods*. The heights of the nodes in all four dendrograms are equal to values of d. The dendrogram produced by unweighted group average clustering is noticeably different from the other three. It does not follow, however, that this will be true with other data matrices.

The unweighted group average method is probably the clustering procedure most widely used by ecologists. To mention only a single example, it

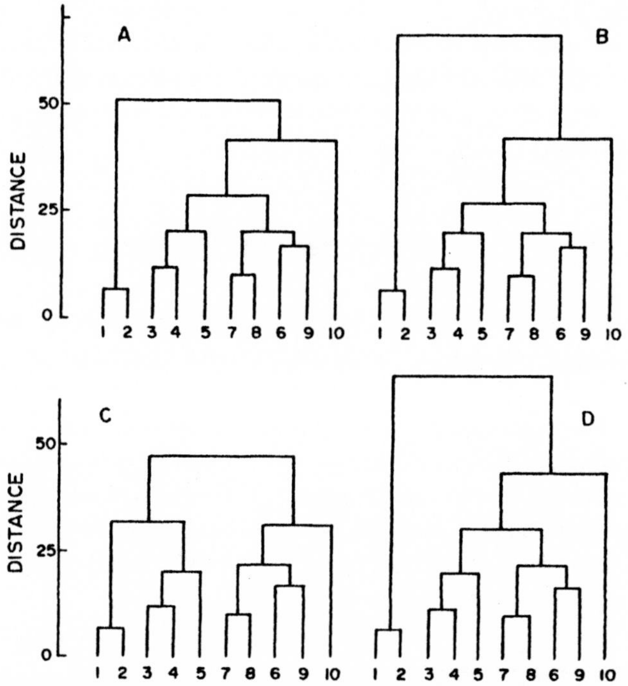

Figure 2.17. Four dendrograms produced by applying different forms of average linkage clustering to Data Matrix #6 (see Table 2.14). (*A*) centroid clustering; (*B*) median clustering; (*C*) unweighted group average clustering; (*D*) weighted group average clustering. The clustering criterion is d^2 for *A* and *B*, and *d* for *C* and *D*.

was used by Strauss (1982) to cluster 43 species of fish occurring in the Susquehanna River drainage of Pennsylvania (this is an example of Q-type clustering; see page 8). As Strauss remarks, "any clustering technique might have been used."

It is, unfortunately, true that no one clustering method is better than all the others in every respect. To choose a method wisely, it is necessary to

TABLE 2.14. DATA MATRIX #6. THE QUANTITIES OF 2 SPECIES IN 10 QUADRATS.

Quadrat	1	2	3	4	5	6	7	8	9	10
Species 1	15	21	33	32	34	51	54	64	66	82
Species 2	75	72	58	46	32	27	45	42	20	15

balance the advantages and disadvantages of each and decide which advantages are most desirable and which disadvantages can be tolerated. The decision is often difficult; choosing the best trade-offs in a given context is always, in the end, somewhat subjective. We now discuss some of the most crucial decisions.

2.8. CHOOSING AMONG CLUSTERING METHODS

Seven clustering techniques have been described in this chapter: nearest and farthest-neighbor clustering, minimum variance clustering, and the four forms of average linkage clustering among which centroid clustering is included. There are many other, less well-known methods, devised for special purposes; accounts of them may be found in more advanced books such as Orlóci (1978), Sneath and Sokal (1973), and Whittaker (1978b). One or another of the last five methods described in this chapter should meet the needs of ecologists in all but exceptional contexts. It remains to compare the methods with one another.

Nearest and Farthest-Neighbor Clustering

These are rarely used nowadays. In these methods, the two clusters to be united at any step are determined entirely by the distance between two individual data points, one in each cluster. Thus a cluster is always represented by only one of its points; moreover, the "representative point" (a different one at each step) is always "extreme" rather than "typical" of the cluster it represents.

Minimum Variance Clustering

This is a useful technique when there is reason to suspect that some (or all) of the quadrats belong to one or more homogeneous classes. For example, suppose data had been collected by sampling, with randomly placed quadrats, a rather heterogeneous tract of forest and scrub. One might be uncertain whether all the quadrats should be thought of as unique or whether, on the contrary, they formed several distinct classes with all the quadrats in any one class constituting a random sample from the same population. In the former case, every node in a clustering dendrogram is interesting and reveals (it is hoped) "true" relationships among dissimilar things. In the latter case the first few fusions do no more than unite groups of quadrats that are not truly distinct from one another; the differences among the quadrats within such a group are due entirely to chance, and the order in which they are united is likewise a matter of chance.

With minimum variance clustering it is possible to do a statistical test of each fusion in order to judge whether the points (or clusters) being united are homogeneous (replicate samples from a single parent population) or heterogeneous (samples from different populations). This is equivalent to judging, objectively, the "information value" of each node in a dendrogram. Thus if the lowermost nodes represent fusions of homogeneous points or clusters, they have no information value; obviously, it is useful to distinguish them from nodes representing the fusions that *do* convey information about the relationships among the clusters and about their relative ecological "closeness." The reader is referred to Goodall (1978b, p. 270) or Orlóci (1978, p. 212) for instructions on how to do the test, which is beyond the scope of this book.

Minimum variance clustering, like farthest-neighbor clustering, tends to give clusters of fairly equal size. If a single data point is equidistant from two cluster centroids and the clusters do not have the same numbers of members, then the data point will unite with the less populous cluster (proved in Orlóci, 1978). The result is that, as clustering proceeds, small clusters acquire new members faster than large ones and chaining is unlikely to happen. This is a great advantage when clustering is done to provide a descriptive classification, for mapping purposes, for instance. Of course, it does not follow that a nicely balanced dendrogram gives a truer picture of ecological relationships than a straggly one.

Average Linkage Clustering

Turning now to the four average linkage clustering techniques, the first choice that must be made is between *unweighted and weighted methods*. In the great majority of cases an unweighted method, which assigns equal weight to each data point and hence weights each cluster according to its size, is better. But if one were studying a mixture of communities and knew that they were very unequally represented in the data, then a weighted method, which assigns equal weight to the clusters irrespective of their sizes, would be useful; it would prevent the abundantly sampled community from having an overly large influence on the shape of the dendrogram. How large is "overly large" is, of course, a question of judgment. Choosing wisely between weighted and unweighted clustering is not always easy, but when in doubt, unweighted clustering is to be preferred.

It remains to choose between *group average clustering and centroid clustering* (or its weighted equivalent, median clustering). The pros and cons are very evenly divided. Each method has a notable advantage that the other lacks and, at the same time, a notable weakness which is a consequence of the advantage.

The strong point of centroid clustering is that each cluster as it is formed is represented by an exactly specifiable point, its centroid, and the distance between two clusters is the distance between their centroids. In group average clustering, there is no such geometrical realism: the clusters cannot be identified with precise representative points and, therefore, the concept of intercluster distance is unavoidably fuzzy. The device of using the average of all interpoint distances between two clusters as a measure of intercluster distance is just that, a device.

The weakness of centroid clustering, a weakness not shared by group average clustering, is that it is not *monotonic*. This term is most easily explained with a figure (Figure 2.18). The upper panel shows six data points and their coordinates in a two-dimensional space. Below are two dendrograms obtained from the data. The dendrogram on the left results from centroid clustering with d^2 as the clustering criterion; the scale shows the square root of the d^2 value corresponding to each node. The dendrogram on the right results from group average clustering with d as criterion.

As may be seen, the centroid clustering dendrogram contains two so-called *reversals*. For example, the height of the node (the intercluster distance) representing the fusion of E with $[D, F]$ is less than that representing the fusion of points D and F. This is because (see the upper panel) although D and F are nearer to each other than either is to E, so that D and F are united first, the centroid of the new cluster after the fusion (the hollow dot labeled $[D, F]$) is nearer to E than either of its component points were before the fusion. The distances are easily found to be

$$d(E, D) = 9.604; \qquad d(E, F) = 9.002; \qquad d(D, F) = 8.944;$$

and

$$d(E, [D, F]) = 8.163.$$

Clearly,

$$d(E, [D, F]) < d(D, F) < \begin{cases} d(E, D) \\ d(E, F) \end{cases}.$$

The reversal where B joins $[A, C]$ has the same cause.

There are no reversals in the dendrogram on the right.

Indeed, it can be proved that reversals cannot occur in a dendrogram obtained by the group average clustering methods (Lance and Williams, 1966), and hence these methods are preferred by those who regard reversals in a dendrogram as a fatal defect.

If a clustering method is incapable of giving reversals, the measure of intercluster distance that it uses is said to be *ultrametric*; with an ultrametric measure, the sequence of intercluster distance values between the pair of clusters united at each successive fusion is always a monotonically (continu-

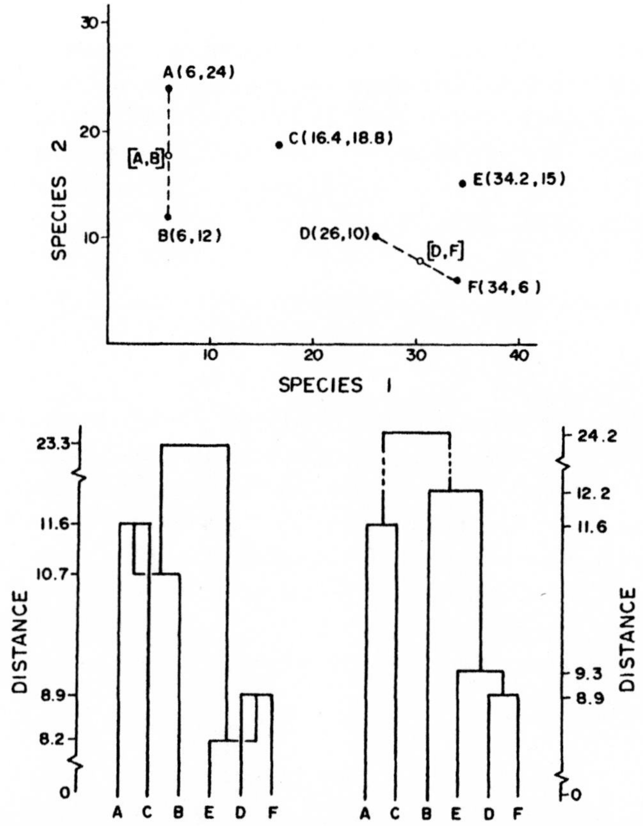

Figure 2.18. Illustration of how reversals appear in a centroid clustering dendrogram (left) although they are absent from the group average clustering dendrogram constructed from the same data (right). The six data points that were clustered are plotted at the top. The hollow dots are the centroids of clusters $[A, B]$ and $[D, F]$. Note the scales of the dendrograms, which have been adjusted to make the reversals conspicuous.

ously) increasing sequence. Therefore, the clustering method is called *monotonic*. In group average clustering the measure of intercluster distance (the average of all the point-to-point distances between a point in one cluster and a point in the other) is ultrametric; hence the method is monotonic and the dendrograms it gives are free of reversals.

If a clustering method can give reversals, the measure of intercluster distance that it uses is not ultrametric and the method is not monotonic. In centroid clustering, the measure of intercluster distance (the distance between the two cluster centroids) is not ultrametric, as Figure 2.18 shows. Hence centroid clustering is nonmonotonic and it can give reversals.

In sum: group average clustering gives clusters with undefined centers and monotonic dendrograms; centroid (including median) clustering gives clusters with exactly defined centers and dendrograms that may contain reversals. It is logically impossible to have the best of both worlds if a guaranteed absence of reversals is indeed "best." Reversals, where they occur, suggest that the difference between the clusters being united is negligible; unfortunately, one cannot make the converse inference, that an absence of reversals implies distinctness of all the clusters.

2.9. RAPID NONHIERARCHICAL CLUSTERING

All the clustering techniques so far described in this chapter have been *hierarchical*. A hierarchical clustering procedure does more than merely unite data points into clusters. It performs the fusions in a definite sequence and, therefore, the outcome can be displayed as a dendrogram, enabling one to discern the different degrees of relationship among the points.

With very large data matrices, it is often desirable to do a nonhierarchical clustering of the data points (quadrats or other sampling units) as a preliminary to hierarchical clustering. There are several reasons for doing this:

1. Hierarchical clustering by any method makes heavy demands on computer time and memory. For very large bodies of data, a computationally more economical procedure is desirable.
2. A dendrogram with a very large number (100 or more, say) of ultimate branches is too big to comprehend.

3. A large data matrix usually contains data from numerous replicate sampling units within each of the distinguishably different communities whose relationships are being investigated. Hence the earliest fusions in a hierarchical clustering are likely to be uninformative. They merely have the effect of pooling replicate quadrats, and the order of the fusions which bring this pooling about is of no interest.

Therefore, it is desirable to subject very large data matrices to nonhierarchical clustering at the outset of an analysis. The clustering should be done by as economical a method, in computational terms, as possible. Also, so far as possible, the clusters it defines should be homogeneous. This preliminary clustering should have the effect of condensing a large data matrix. It should permit batches (or "pools") of replicate quadrat records in the large matrix to be replaced by the average for each pool. Then the centroids of these batches, or pools, of virtually indistinguishable quadrats can become data points for a hierarchical clustering that will reveal their relationships.

It is unfortunate that the word *clustering* is at present used both for the hierarchical clustering procedures discussed in earlier sections of this chapter, and also for the rapid, preliminary nonhierarchical clustering that we are considering now. The objectives of the two operations are entirely different. Nonhierarchical clustering is just a way of boiling down unmanageably large data matrices in order to make hierarchical clustering (or other analyses) computationally feasible and ecologically informative. To avoid ambiguity, there should be different names for the two operations; perhaps rapid nonhierarchical clustering could be called *pooling*, since that is what it does, and a cluster defined by such a process could be called a *pool*, as in the preceding paragraph. These newly coined terms are used in what follows.

Several methods of data pooling have been devised; probably the best is Gauch's (1980, 1982a) technique which he calls "composite clustering." A computer program for doing it is available (Gauch, 1979). In outline, the process is as follows.

There are two phases. In the first phase, points are selected at random from the swarm of data points, and all other points within a specified radius of each selected point are assigned to a pool centered on that point. The random points that act as pool centers (they are not, of course, centroids) are chosen one after another. The earliest pools are hyperspherical in shape (circular in the two-dimensional case). Later pools are not allowed to overlap earlier pools (i.e., a point must remain a member of the pool it joins first) and hence these later pools tend to be small and "spikey" since they occupy the interstices among earlier formed pools. Therefore, the procedure has a second phase in which pools with fewer than a specified quota of points are broken up. Their member points are reassigned to the nearest large pool, provided that they lie within a predetermined distance of it. Thus on the second round the radii of some pools are slightly increased. The number of pools formed is under the control of the investigator, who must choose the radius to be used at each phase of the process. The smaller the radii, the smaller and more numerous the pools, and the more confident one can be that they are homogeneous. Points that fail to become numbers of any pool are rejected as "outliers." After the pooling has been completed, each pool can be replaced by an average point (the centroid of all the points in the pool) and these average points then become the data points for further investigations.

APPENDIX

Apollonius's Theorem

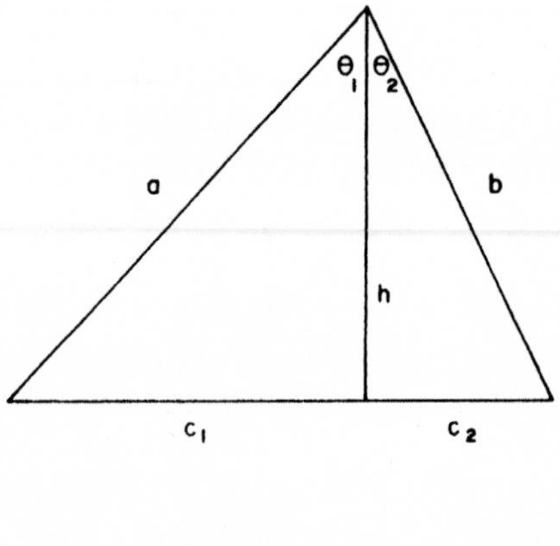

To Prove:

$$c^2 = a^2 + b^2 - 2ab\cos\theta.$$

where

$$c = c_1 + c_2 \quad \text{and} \quad \theta = \theta_1 + \theta_2.$$

Proof: First recall that

$$\cos(\theta_1 + \theta_2) = \cos\theta_1\cos\theta_2 - \sin\theta_1\sin\theta_2.$$

Observe, from the figure, that

$$h = a\cos\theta_1 = b\cos\theta_2; \qquad \therefore \quad h^2 = ab\cos\theta_1\cos\theta_2;$$

$$c_1 = a\sin\theta_1 \quad \text{and} \quad c_2 = b\sin\theta_2.$$

Also,

$$c_1^2 = a^2 - h^2 \quad \text{and} \quad c_2^2 = b^2 - h^2.$$

Now

$$c^2 = (c_1 + c_2)^2 = c_1^2 + c_2^2 + 2c_1c_2.$$

$$\begin{aligned}
\therefore \quad c^2 &= (a^2 - h^2) + (b^2 - h^2) + 2ab\sin\theta_1\sin\theta_2 \\
&= a^2 + b^2 - 2(h^2 - ab\sin\theta_1\sin\theta_2) \\
&= a^2 + b^2 - 2ab(\cos\theta_1\cos\theta_2 - \sin\theta_1\sin\theta_2) \\
&= a^2 + b^2 - 2ab\cos\theta.
\end{aligned}$$

QED

EXERCISES

2.1. Given the following data matrix **X**, what is the Euclidean distance between: (a) points 1 and 5; (b) points 2 and 3; (c) points 3 and 5? (Each point is represented by a column of **X**, which gives its coordinates in four-space.)

$$\mathbf{X} = \begin{pmatrix} 3 & 8 & 2 & -1 & 0 \\ 0 & 0 & 4 & -2 & 3 \\ 1 & 3 & -1 & 6 & 3 \\ -2 & -4 & -2 & 0 & 7 \end{pmatrix}.$$

2.2. Cluster the five quadrats in Exercise 2.1 using farthest-neighbor clustering. Tabulate the results in a table like Table 2.5.

2.3. Suppose the data points whose coordinates are given by the columns of **X** in Exercise 2.1 were assigned to two classes: [1,2] and [3,4,5]. Find the coordinates of the centroid of each of these classes. What is the distance between the two centroids?

2.4. Let M, N, and P denote the centroids of clusters of points in five-space with, respectively, $m = 5$, $n = 15$, and $p = 6$ members. Find the distance^2 between P and the centroid of the cluster formed by uniting clusters $[M]$ and $[N]$. The coordinates of points M, N, and P are as follows:

$$\begin{array}{ccc} \text{M} & \text{N} & \text{P} \end{array}$$
$$\begin{pmatrix} 1 & -5 & 10 \\ -8 & 2 & 11 \\ 8 & 5 & -2 \\ 9 & 1 & -5 \\ 7 & 4 & -6 \end{pmatrix}$$

2.5. What is the within-cluster dispersion of the swarm of five points whose coordinates are given by **X** in Exercise 2.1?

2.6. For the two points in six-space whose coordinates are given by the columns of the following 6×2 matrix, find: (a) the chord distance; (b) the geodesic metric; (c) the angular separation between the two points.

$$\begin{pmatrix} 3 & -1 \\ 4 & -3 \\ -2 & -4 \\ 1 & 0 \\ 5 & 4 \end{pmatrix}$$

2.7. Obtain Jaccard's and Sørensen's indices of similarity (J and S) for the following three pairs of quadrats in Data Matrix #5 (Table 2.13, page 62): (a) quadrats 1 and 2; (b) quadrats 3 and 4; (c) quadrats 2 and 3. Prove that S must always exceed J except when $S = J = 1$.

2.8. The columns of the following matrix give the coordinates in four-space of seven points grouped into clusters as shown.

$$\overbrace{\underbrace{}_{}\!\!\overbrace{M_1 \quad M_2}^{[M]} \quad \overbrace{N_1 \quad N_2 \quad N_3}^{[N]}}^{[Q]} \quad \overbrace{P_1 \quad P_2}^{[P]}$$

$$\begin{pmatrix} 3 & 3 & 8 & 9 & 7 & -3 & -1 \\ 2 & 3 & 7 & 5 & 5 & -2 & -1 \\ 1 & -1 & 8 & 8 & 6 & 0 & 4 \\ 4 & 0 & 6 & 9 & 6 & 1 & -2 \end{pmatrix}$$

Find the following measures of the dissimilarity between clusters $[P]$ and $[Q]$: (a) The unweighted average distance; (b) the centroid distance; (c) the weighted average distance; (d) the median distance.

Chapter Three

Transforming Data Matrices

3.1. INTRODUCTION

This chapter provides an elementary introduction to the mathematics necessary for an understanding of the ordination techniques described in Chapter 4. But to begin, it is desirable to demonstrate a crude form of ordination to show what the purpose of ordination is and how this purpose is achieved.

Consider Data Matrix #7 (Table 3.1) which shows the quantities of four species in six quadrats (or other sampling units). Suppose one were asked to list the quadrats "in order" or, equivalently, to rank them. Clearly, there is no "natural" way to do this; the data points do not have any intrinsic order. The task would be simple if one species only had been recorded; then the quadrats could be ranked in order of increasing (or decreasing) quantity of the single species. When two or more species are recorded for each quadrat, however, the data points do not, usually, fall in a natural sequence. Although such natural ordering is not logically impossible, in practice one is far more likely to find that a set of observed data points represents a diffuse swarm in a space of many dimensions. Therefore, if one wishes to rank the points, it is necessary first to prescribe some method of assigning a single numerical score to each quadrat. Then, and only then, can the points (quadrats) be ranked, using the scores to decide the ranking.

To illustrate, suppose the quantities in Data Matrix #7 are cover values of four species of forest plants. Let species 1 be a canopy tree, species 2 a subdominant tree, species 3 a tall shrub, and species 4 a low shrub. One way to assign a score to each quadrat would be simply to add the cover values of all four species. Thus letting x_{ij} denote the cover of species i in quadrat j, the score of quadrat j is $(x_{1j} + x_{2j} + x_{3j} + x_{4j})$. Using this scoring method, the scores of quadrats 1 through 6 are found to be, respectively,

118 183 83 150 130 124;

the ranking of the quadrats, from that with the smallest to that with the largest score, is then

#3, #1, #6, #5, #4, #2.

Alternatively, one might choose to weight the species according to their sizes instead of treating them all equally. There are infinitely many ways in which this could be done. For example, one might assign to quadrat j the score $(4x_{1j} + 3x_{2j} + 2x_{3j} + x_{4j})$. Using this formula, the scores of quadrats 1 through 6 are, respectively,

335 340 221 400 234 375

and the ranking of the quadrats becomes

#3, #5, #1, #2, #6, #4.

This is a different order from that given in the preceding paragraph. Both lists are *ordinations* of the data in Data Matrix #7 and the fact that they are different shows that the result of an ordination depends on the method chosen for assigning scores to the quadrats or, equivalently, on the weights assigned to the different species. The various ordination techniques de-

TABLE 3.1. DATA MATRIX #7. THE QUANTITIES OF $s = 4$ SPECIES IN $n = 6$ QUADRATS.

Quadrat	1	2	3	4	5	6
Species 1	50	20	25	45	15	60
Species 2	11	16	20	33	14	17
Species 3	45	65	23	49	31	37
Species 4	12	82	15	23	70	10

scribed in Chapter 4 are all procedures for determining these weights objectively instead of choosing them arbitrarily and subjectively as was done in the preceding.

3.2. VECTOR AND MATRIX MULTIPLICATION

The operation just performed, that of transforming a data matrix to a list of scores that can be ranked, can be represented symbolically.

Vector × Matrix Multiplication

Let us write **X** for the data matrix which in the previous example is an array of numbers (*elements*) arranged in four rows and six columns enclosed in large parentheses. It is a 4 × 6 matrix.

$$\mathbf{X} = \begin{pmatrix} x_{11} & x_{12} & x_{13} & x_{14} & x_{15} & x_{16} \\ x_{21} & x_{22} & x_{23} & x_{24} & x_{25} & x_{26} \\ x_{31} & x_{32} & x_{33} & x_{34} & x_{35} & x_{36} \\ x_{41} & x_{42} & x_{43} & x_{44} & x_{45} & x_{46} \end{pmatrix}.$$

The element in the ith row and jth column is written x_{ij}. Notice that the first subscript in x_{ij} is the number of the row in which the element appears, and the second subscript is the number of the column; this rule is invariable and is adhered to by all writers. In this book, and in most but not all ecological writing, data matrices are so arranged that the rows represent species and the columns represent sampling plots or quadrats. Therefore, when this system is used, x_{ij} means the amount of species i in quadrat j. The single symbol **X** denotes the whole matrix, made up in this case of 24 distinct numbers. It does not denote a single number (a *scalar*). Boldface type is used for **X** to show that it is a matrix, not a scalar.

Now let us write $\mathbf{y}'$ for the list of six scores; $\mathbf{y}'$ is a matrix with only one row, otherwise known as a *row vector*. It is

$$\mathbf{y}' = (y_1, y_2, y_3, y_4, y_5, y_6).$$

As before, the boldface type shows that $\mathbf{y}'$ is a matrix. The lowercase letter (**y** not **Y**) shows that it is a *vector* (a matrix with only one row, or only one column). The prime shows that it is a row vector. If the same array of elements were written as a column instead of a row, they would form a *column vector* and be denoted by **y** (without a prime).

Finally, let us write $\mathbf{u}'$ for the list of coefficients by which each element in a column of **X** is to be multiplied to yield an element of $\mathbf{y}'$; $\mathbf{u}'$ is a row vector and the number of elements it contains must obviously be the same as the number of elements in a column of **X**. (The number of elements in a column of **X** is, of course, the number of rows of **X**.) Hence $\mathbf{u}' = (u_1, u_2, u_3, u_4)$. Recall again the example in Section 3.1. The first list of quadrat scores, namely,

$$\mathbf{y}' = (118, 183, 83, 150, 130, 124)$$

was obtained by adding the elements in each column of **X**. That is, the score for the jth quadrat was given by

$$y_j = x_{1j} + x_{2j} + x_{3j} + x_{4j} = \sum_{i=1}^{4} x_{ij}.$$

Alternatively, this can be written as

$$y_j = u_1 x_{1j} + u_2 x_{2j} + u_3 x_{3j} + u_4 x_{4j} = \sum_{i=1}^{4} u_i x_{ij},$$

with $u_1 = u_2 = u_3 = u_4 = 1$. Therefore, these scores were obtained using the row vector

$$\mathbf{u}' = (1, 1, 1, 1).$$

The elements in the second list of scores, namely,

$$\mathbf{y}' = (335, 340, 221, 400, 234, 375)$$

were obtained from the same formula ($y_j = \Sigma_{i=1}^{4} u_i x_{ij}$) but with $u_1 = 4$, $u_2 = 3$, $u_3 = 2$, and $u_4 = 1$. Hence in the second case we had

$$\mathbf{u}' = (4, 3, 2, 1).$$

The operation by which $\mathbf{y}'$ was obtained from $\mathbf{u}'$ and **X** in the two cases is a form of *matrix multiplication*. It entails the multiplication of a matrix by a row vector (or one-row matrix). Thus $\mathbf{y}'$ is the product of $\mathbf{u}'$ and $\mathbf{X}$. Written as an equation, this is

$$\mathbf{u}'\mathbf{X} = \mathbf{y}'.$$

In words: the coefficient vector $\mathbf{u}'$ times the data matrix $\mathbf{X}$ is the score vector $\mathbf{y}'$. This is identical in meaning to the much clumsier equation

$$(u_1, u_2, u_3, u_4)\begin{pmatrix} x_{11} & x_{12} & x_{13} & x_{14} & x_{15} & x_{16} \\ x_{21} & x_{22} & x_{23} & x_{24} & x_{25} & x_{26} \\ x_{31} & x_{32} & x_{33} & x_{34} & x_{35} & x_{36} \\ x_{41} & x_{42} & x_{43} & x_{44} & x_{45} & x_{46} \end{pmatrix}$$

$$= (y_1, y_2, y_3, y_4, y_5, y_6).$$

This extended version of the equation $\mathbf{u}'\mathbf{X} = \mathbf{y}'$ is itself a condensed form of six separate equations, of which the first and last are:

$$u_1x_{11} + u_2x_{21} + u_3x_{31} + u_4x_{41} = y_1$$

$$\cdots\cdots\cdots\cdots\cdots\cdots\cdots\cdots$$

$$u_1x_{16} + u_2x_{26} + u_3x_{36} + u_4x_{46} = y_6.$$

Thus the rule for calculating each of the six elements of $\mathbf{y}'$, that is, for calculating the elements in the product $\mathbf{u}'\mathbf{X}$, is the formula already given:

$$y_j = \sum_{i=1}^{4} u_i x_{ij} \qquad \text{for } j = 1, 2, \ldots, 6.$$

To generalize: suppose an $s \times n$ data matrix $\mathbf{X}$ records the amounts of s species in n quadrats. Let $\mathbf{u}'$ be an s-element row vector (i.e., a $1 \times s$ matrix) of weighting coefficients; these are the weights to be assigned to each species in order to calculate the score for a quadrat. Let the resultant scores be listed in the n-element row vector $\mathbf{y}'$. Then

$$\mathbf{u}'\mathbf{X} = \mathbf{y}' \tag{3.1}$$

in which

$$y_j = \sum_{i=1}^{s} u_i x_{ij} \qquad \text{for } j = 1, 2, \ldots, n.$$

Let us rewrite Equation (3.1) with the sizes of the three matrices shown below them:

$$\underset{(1\times s)}{\mathbf{u}'}\ \underset{(s\times n)}{\mathbf{X}} = \underset{(1\times n)}{\mathbf{y}'}\ .$$

For the multiplication to be possible, the number of columns in the first factor, $\mathbf{u}'$, *must* be the same as the number of rows in the second factor, $\mathbf{X}$. Since $\mathbf{u}'$ has s columns and $\mathbf{X}$ has s rows, the product $\mathbf{y}' = \mathbf{u}'\mathbf{X}$ can indeed be formed. It has the same number of rows as the first factor, $\mathbf{u}'$, and the same number of columns as the second factor, $\mathbf{X}$. In other words, the size of $\mathbf{y}'$ is $1 \times n$.

As should now be clear, the factors in a matrix product must appear in correct order. Equation (3.1) can*not* be written as $\mathbf{X}\mathbf{u}' = \mathbf{y}'$. The product $\mathbf{X}\mathbf{u}'$ does not exist, since n, the number of columns in $\mathbf{X}$, is not equal to 1, the number of rows in $\mathbf{u}'$.

The product $\mathbf{u}'\mathbf{X}$ is described as $\mathbf{X}$ *premultiplied* by $\mathbf{u}'$ or as $\mathbf{u}'$ *postmultiplied* by $\mathbf{X}$.

Matrix × Matrix Multiplication

The preceding paragraphs showed how to premultiply a data matrix $\mathbf{X}$ by a vector $\mathbf{u}'$ of weighting coefficients to obtain a vector $\mathbf{y}'$ of quadrat scores. Before proceeding, it is worthwhile to recall the purpose of the operation. It is to replace a large, perhaps confusing, data matrix by a list of scores that is much more easily comprehended. To put the argument in geometric terms: an $s \times n$ data matrix is equivalent to a swarm of n points in s-dimensional space. Therefore, unless $s \leq 3$, the swarm is impossible to visualize. However, if the original data matrix is transformed into a list of "quadrat scores" by the procedure previously described, the multidimensional swarm of points is transformed into a one-dimensional row of points that can easily be plotted on one axis to make a one-dimensional graph.

Reducing an s-dimensional swarm to a one-dimensional row entails considerable sacrifice of information, of course. This raises the question: Need multidimensional data be so severely condensed to make them comprehensible? The answer is obviously no. A two-dimensional swarm (a conventional scatter diagram) is quite as easy to understand; it can be plotted on a sheet of paper. How, then, can the original s-dimensional swarm be transformed to a two-dimensional swarm?

An obvious way is to carry out the described procedure twice over using two different vectors of weighting coefficients, $\mathbf{u}_1'$ and $\mathbf{u}_2'$, say. Two vectors, $\mathbf{y}_1'$ and $\mathbf{y}_2'$, of scores are obtained, each with n elements. Thus each of the n points now has two scores which can be treated as the coordinates of a point in two-dimensional space enabling the data to be plotted as an ordinary scatter diagram.

To illustrate, consider Data Matrix #7 again. It has already been condensed to a one-dimensional list of scores in two different ways. The first condensation used the vector $(1, 1, 1, 1) = \mathbf{u}_1'$, and gave the result $(118, 183, 83, 150, 130, 124) = \mathbf{y}_1'$. The second used the vector $(4, 3, 2, 1) = \mathbf{u}_2'$ and gave the result $(335, 340, 221, 400, 234, 375) = \mathbf{y}_2'$. It is straightforward to combine these two sets of results. We let quadrat 1 be represented by the pair of scores (118, 335), quadrat 2 by the pair of scores (183, 340), and so on. Each pair of scores is treated as a pair of coordinates and the points are plotted in a two-dimensional coordinate frame, with the first scores measured along the abscissa and the second scores along the ordinate. The result, a two-dimensional ordination, is shown in Figure 3.1.

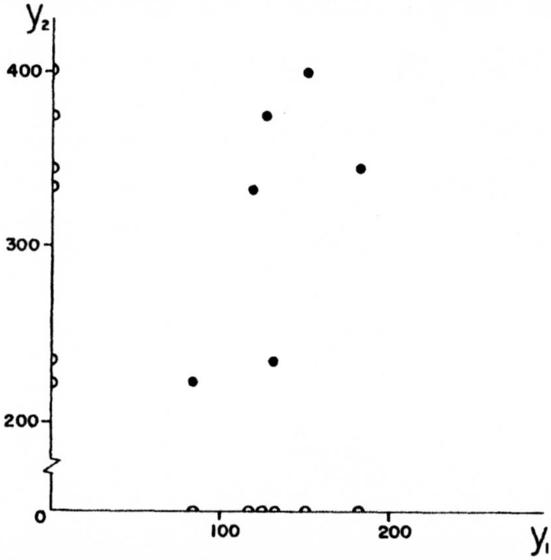

Figure 3.1. The data points of Data Matrix #7 after the transformation of their original coordinates in four-space, given in Table 3.1, to coordinates in two-space. The solid dots show the two-dimensional ordination of the original data described in the text. The hollow half-dots on the y_1 and y_2 axes show the two one-dimensional ordinations of the data yielded by vectors $\mathbf{u}_1'$ and $\mathbf{u}_2'$, respectively.

The two operations just performed on data matrix **X** could be represented symbolically by the two equations

$$\mathbf{u}_1'\mathbf{X} = \mathbf{y}_1' \quad \text{and} \quad \mathbf{u}_2'\mathbf{X} = \mathbf{y}_2'.$$

However, there is a still more compact representation, namely,

$$\mathbf{UX} = \mathbf{Y} \tag{3.2}$$

Here **U** has two rows, the first being $\mathbf{u}_1'$ and the second $\mathbf{u}_2'$. That is, **U** is the 2 × 4 matrix

$$\mathbf{U} = \begin{pmatrix} 1 & 1 & 1 & 1 \\ 4 & 3 & 2 & 1 \end{pmatrix} = \begin{pmatrix} u_{11} & u_{12} & u_{13} & u_{14} \\ u_{21} & u_{22} & u_{23} & u_{24} \end{pmatrix}.$$

Notice that the 2 × 4 matrix **U** is denoted by a capital letter since lowercase letters are reserved for vectors. Also, the elements of **U** now require a pair of subscripts to define their locations in the matrix; the first subscript specifies the row and the second the column.

Likewise, matrix **Y** in Equation (3.2) is a matrix with two rows and six columns. It is

$$\mathbf{Y} = \begin{pmatrix} 118 & 183 & 83 & 150 & 130 & 124 \\ 335 & 340 & 221 & 400 & 234 & 375 \end{pmatrix}$$

$$= \begin{pmatrix} y_{11} & y_{12} & y_{13} & y_{14} & y_{15} & y_{16} \\ y_{21} & y_{22} & y_{23} & y_{24} & y_{25} & y_{26} \end{pmatrix}.$$

To generalize: Equation (3.2) specifies that **X** is to be premultiplied by **U** to give **Y**. Suppose data matrix **X** were of size $s \times n$. Then writing (3.2) with the dimensions of each matrix shown gives

$$\underset{(2\times s)}{\mathbf{U}}\ \underset{(s\times n)}{\mathbf{X}} = \underset{(2\times n)}{\mathbf{Y}}.$$

The factors are so ordered (**UX**, not **XU**) that the number of columns in the first factor is equal to the number of rows in the second; both are s. The product **Y** has the same number of rows as the first factor and the same number of columns as the second factor.

We can generalize further. So far we have discussed one-dimensional ordination and two-dimensional ordination. There is no need to stop at two dimensions. It is true that a three-dimensional ordination yields a three-dimensional swarm of points that can only be plotted on paper as a perspective figure or as three separate two-dimensional graphs; higher-dimensional ordinations require even more two-dimensional graphs for their portrayal. However, the process of ordinating a multidimensional swarm of data points obviously entails a trade-off. One must balance the advantages of condensing the data against the disadvantages of sacrificing information. It is often desirable to keep more than three dimensions in transformed data, equivalently, to do a more-than-three-dimensional ordination. The topic is discussed in Chapter 4. For the present, let us consider the symbolic representation of a p-dimensional ordination of an $s \times n$ data matrix **X**. The required representation has already been given: it is Equation (3.2), unchanged. A difference appears only if the sizes of the matrices are shown. We then have

$$\underset{(p\times s)}{\mathbf{U}}\ \underset{(s\times n)}{\mathbf{X}} = \underset{(p\times n)}{\mathbf{Y}}. \tag{3.3}$$

Each of the p rows of **U** is a set of s weighting coefficients (how to find numerical values for these coefficients, objectively, is considered in Chapter 4). Each of the n columns of **Y** is a set of p scores for one of the quadrats; treating these scores as coordinates permits the points to be plotted (conceptually) in a p-dimensional coordinate frame.

The rule for calculating the value of the (i, j)th element of **Y**, that is, the ith score of the jth point, is summed up in the equation

$$y_{ij} = u_{i1}x_{1j} + u_{i2}x_{2j} + \cdots + u_{is}x_{sj} = \sum_{r=1}^{s} u_{ir}x_{rj}.$$

In words, the (i, j)th element of **Y** is the sum of the pairwise products of the elements in the ith row of the first factor, **U**, and the jth column of the second factor, **X**.

Equivalently, as was just done, one can think of each of the p rows of **U** as a row vector $\mathbf{u}'$, and the corresponding row of **Y** as a row vector $\mathbf{y}'$. One then performs the multiplication $\mathbf{u}'\mathbf{X} = \mathbf{y}'$, p separate times. Finally, the p vectors $\mathbf{y}'$, each with n elements, are stacked on top of one another to give the $p \times n$ matrix **Y**.

Linear Transformations

In what follows, a matrix of size $s \times n$, that is, with s rows and n columns, is called an $s \times n$ matrix.

It was shown in Equation (3.3) that when an $s \times n$ data matrix **X** is premultiplied by a $p \times s$ matrix **U**, the product **Y** is a $p \times n$ matrix. Now matrix **X** specifies the locations of n points in s-dimensional space (s-space, for short). Indeed, each column of **X** is a list of the s coordinates of one of the points. Likewise, **Y** specifies the locations of n points in p-space; each of its columns is a list of the p coordinates of one of the points.

We can, therefore, regard **Y** as an altered form of **X**. Both matrices amount to instructions for mapping the same swarm of n points. **X** maps them in s-space; **Y** maps the same points in p-space. Therefore, if $p < s$, the p-dimensional swarm of points whose coordinates are given by the columns of **Y** is a "compressed" version of the original s-dimensional swarm of points whose coordinates were given by the columns of **X**. In other words, premultiplying **X** by the $p \times s$ matrix **U** has the effect of condensing the data and, inevitably, of obliterating some of the information the original data contained.

Now suppose that $p = s$ or, equivalently, that **U** is an $s \times s$ matrix (a square matrix). Premultiplying **X** by **U** no longer condenses the data since the product **Y** is, like the original **X**, an $s \times n$ matrix. But the premultiplication does affect the shape of the swarm represented by **X**, and it is interesting to see how a very simple swarm is affected, geometrically, by a variety of different versions of the "transforming" matrix **U**.

To make the demonstration as clear as possible, we shall put

$$\mathbf{X} = \begin{pmatrix} 1 & 10 & 1 & 10 \\ 1 & 1 & 10 & 10 \end{pmatrix}.$$

Thus **X** is a 2×4 matrix representing a swarm of $n = 4$ points in $s =$ two-space. The points are at the corners of a square (see Figure 3.2*a*).

We now evaluate **UX** using several different **U**s. The numerical equations are given in the following; **X** is written out in full only in the first equation and is left as the symbol **X** subsequently. The first factor in each equation is the matrix **U** whose effect is being examined. The results are plotted in Figure 3.2. All the transformations illustrated have their counterparts in spaces of more than two dimensions, of course, but these are difficult (when

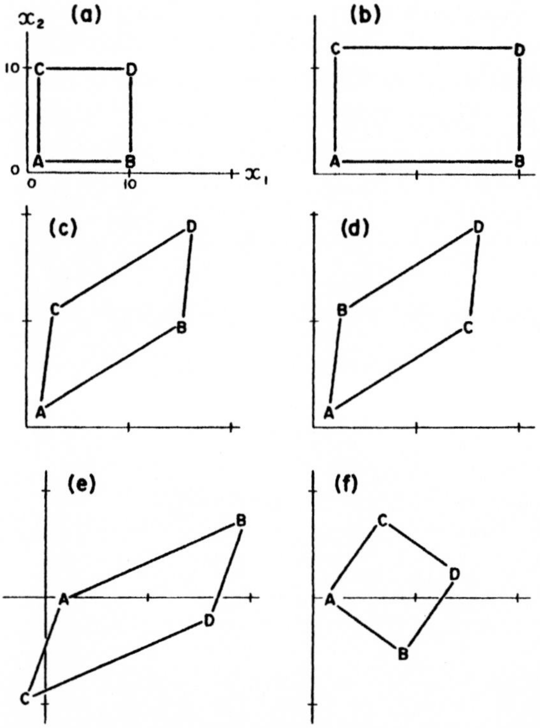

Figure 3.2. (a) The four data points of the matrix **X** and also of **IX** = **X** (see text). (b)–(f). The same data points after transformation by the five different 2 × 2 matrices given in the text. The lines joining points A, B, C, and D have been drawn to emphasize the shape of the "swarm" of four points.

$s = 3$) or impossible (when $s > 3$) to draw. The reader should experiment with other **U**s as well.

(a)
$$\begin{pmatrix} 1 & 0 \\ 0 & 1 \end{pmatrix}\begin{pmatrix} 1 & 10 & 1 & 10 \\ 1 & 1 & 10 & 10 \end{pmatrix} = \begin{pmatrix} 1 & 10 & 1 & 10 \\ 1 & 1 & 10 & 10 \end{pmatrix}.$$

Here **U** is the so-called *identity matrix*, always denoted by **I**, which has 1s on the main diagonal (top left to bottom right) and 0s elsewhere. In symbols the equation is, therefore, **IX** = **X**. It is apparent that premultiplication of **X** by **I** leaves **X**, and the square it represents, unchanged.

(b)
$$\begin{pmatrix} 2 & 0 \\ 0 & 1.2 \end{pmatrix}\mathbf{X} = \begin{pmatrix} 2.0 & 20.0 & 2 & 20 \\ 1.2 & 1.2 & 12 & 12 \end{pmatrix}.$$

This $\mathbf{U}$ is a *diagonal matrix*; that is, its only nonzero elements are those on the main diagonal. It is seen that each row of $\mathbf{Y} = \mathbf{UX}$ is the corresponding row of $\mathbf{X}$ multiplied by the single nonzero element in the same row of $\mathbf{U}$. The geometrical effect is to change the scales of the two axes, and the original square becomes a rectangle. Obviously, if we had put

$$\mathbf{U} = \begin{pmatrix} 3 & 0 \\ 0 & 3 \end{pmatrix},$$

the original square would have remained a square but with sides three times as long.

$$(c) \qquad \begin{pmatrix} 1.5 & 0.1 \\ 0.9 & 1.0 \end{pmatrix}\mathbf{X} = \begin{pmatrix} 1.6 & 15.1 & 2.5 & 16 \\ 1.9 & 10.0 & 10.9 & 19 \end{pmatrix}.$$

The square is now transformed to a parallelogram.

$$(d) \qquad \begin{pmatrix} 0.1 & 1.5 \\ 1.0 & 0.9 \end{pmatrix}\mathbf{X} = \begin{pmatrix} 1.6 & 2.5 & 15.1 & 16 \\ 1.9 & 10.9 & 10.0 & 19 \end{pmatrix}.$$

The parallelogram is of the same shape as in (c) but the corners B and C are interchanged.

$$(e) \qquad \begin{pmatrix} 2.0 & -0.4 \\ 0.8 & -1.0 \end{pmatrix}\mathbf{X} = \begin{pmatrix} 1.6 & 19.6 & -2.0 & 16 \\ -0.2 & 7.0 & -9.2 & -2 \end{pmatrix}.$$

There is no reason why all the elements of $\mathbf{U}$ or all the coordinates of the data points should be positive; although no measured species quantities are negative, of course, it is often desirable to convert these measurements to deviations from their mean values, as shown in Section 3.4, and when this is done some elements of $\mathbf{X}$ must be negative. This example illustrates the effect of setting some of the elements of $\mathbf{U}$ negative.

$$(f) \qquad \begin{pmatrix} 0.8 & 0.6 \\ -0.6 & 0.8 \end{pmatrix}\mathbf{X} = \begin{pmatrix} 1.4 & 8.6 & 6.8 & 14 \\ 0.2 & -5.2 & 7.4 & 2 \end{pmatrix}.$$

The original square is still a square and its size is unchanged, but it has been rotated. A matrix $\mathbf{U}$ that has this effect is known as *orthogonal*. Because of their great importance in data transformations, orthogonal matrices require detailed description in the following.

First, however, a word on terminology. All the operations on **X** previously illustrated and others like them in which **U** is an $s \times s$ matrix are known as *linear transformations* of **X**. The word *linear* implies that each element in **Y** is a linear function of the elements of **X**, that is, one in which the elements are multiplied by constants and added, but are not multiplied by each other or squared or raised to higher powers. In other words, all the xs are said to be in the first degree.

For instance, recall that

$$y_{ij} = u_{i1}x_{1j} + u_{i2}x_{2j} + \cdots + u_{is}x_{sj}.$$

In this equation, which expresses y_{ij} as a function of $x_{1j}, x_{2j}, \ldots, x_{sj}$, the factors $u_{i1}, u_{i2}, \ldots, u_{is}$ are constants. They are independent of j.

Now suppose that the original data had been only one-dimensional, that is, that only one species had been observed so that $s = 1$. Then the transformation equation

$$\underset{(s\times s)}{\mathbf{U}}\ \underset{(s\times n)}{\mathbf{X}} = \underset{(s\times n)}{\mathbf{Y}}$$

would be reduced to

$$u\mathbf{x}' = \mathbf{y}'$$

where u is a scalar (an ordinary number) and $\mathbf{x}'$ and $\mathbf{y}'$ are both n-element row vectors (equivalently, $1 \times n$ matrices).

Written out *in extenso*, the last equation is

$$u(x_1, x_2, \ldots, x_n) = (y_1, y_2, \ldots, y_n).$$

To multiply a vector by a scalar, one simply multiplies each separate element of the vector by the scalar. Thus the equation $u\mathbf{x}' = \mathbf{y}'$ is a condensed form of the n separate equations

$$ux_1 = y_1; \qquad ux_2 = y_2; \qquad \ldots; \qquad ux_n = y_n.$$

Each of these is the equation of a straight line. Hence the adjective *linear* to describe the transformation.

Orthogonal Matrices and Rigid Rotations

It was already mentioned that the 2×2 matrix

$$\mathbf{U} = \begin{pmatrix} 0.8 & 0.6 \\ -0.6 & 0.8 \end{pmatrix}$$

is described as orthogonal. We saw that the transformed data matrix $\mathbf{Y} = \mathbf{UX}$ specifies a swarm of points with the same pattern as that specified by the original $\mathbf{X}$; the only change brought about by the transformation is that the swarm as a whole has a new position relative to the axes of the coordinate frame.

We are, therefore, free to regard the transformation as a movement of the swarm relative to the axes, or of the axes relative to the swarm (see Figure 3.3). In both cases, as the figure shows, the movement consists of a rigid rotation around the origin of the coordinates. In Figure 3.3*b* the swarm of points, behaving as a rigid, undeformable unit, has rotated clockwise around the origin. In Figure 3.3*c* the axes have rotated counterclockwise around the origin, relative to the swarm.

We now enquire: How can a matrix $\mathbf{U}$ be constructed so that its only effect on $\mathbf{X}$ is to cause a rigid rotation of the data swarm relative to the coordinate axes or vice versa?

To answer the question, envisage a single datum point in two-space with coordinates (x_1, x_2). The data matrix is, therefore, the 2×1 matrix (or two-element column vector)

$$\mathbf{x} = \begin{pmatrix} x_1 \\ x_2 \end{pmatrix}$$

(Notice that since $\mathbf{x}$ denotes a column vector it is printed as a lowercase boldface letter without a prime).

Next suppose the axes are rotated counterclockwise around the origin through angle θ. Let the coordinates of the datum point relative to the new, rotated axes be given by

$$\mathbf{y} = \begin{pmatrix} y_1 \\ y_2 \end{pmatrix}$$

(see Figure 3.4).

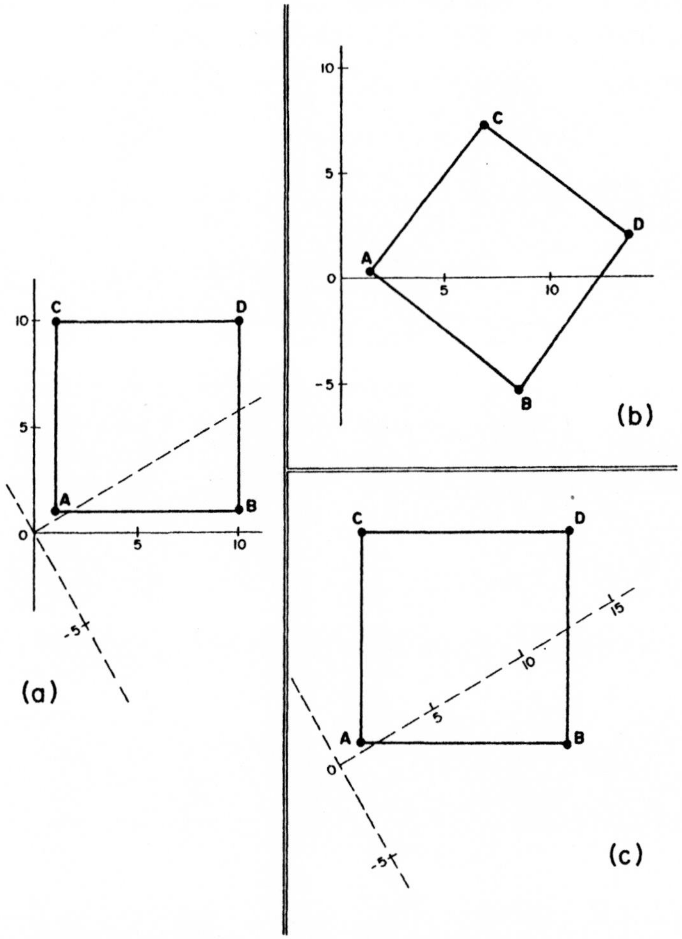

Figure 3.3. (*a*) Relative to the axes shown as solid lines, the points A, B, C, and D have coordinates given by the columns of

$$\mathbf{X} = \begin{pmatrix} 1 & 10 & 1 & 10 \\ 1 & 1 & 10 & 10 \end{pmatrix}.$$

Relative to the dashed axes the coordinates are

$$\mathbf{Y} = \begin{pmatrix} 1.4 & 8.6 & 6.8 & 14 \\ 0.2 & -5.2 & 7.4 & 2 \end{pmatrix}.$$

(*b*) and (*c*) show the transformation of **X** to **Y** in two different ways: (*b*) shows the axes unaltered but the points moved; (*c*) shows the points unaltered but the axes moved.

We need to find y_1 and y_2 in terms of x_1, x_2, and θ from straightforward geometrical and trigonometrical considerations.

Consider Figure 3.4. The points are labeled so that

$$\text{OA} = \text{BP} = x_1; \quad \text{OB} = \text{AP} = x_2; \quad \text{OR} = \text{SP} = y_1; \quad \text{OS} = \text{RP} = y_2.$$

AM is the perpendicular from A to the y_1-axis.

Obviously, OR = OM + MR.
It is seen that

$$\text{OM} = \text{OA}\cos\theta = x_1 \cos\theta,$$

and

$$\begin{aligned} \text{MR} &= \text{MN} + \text{NR} \\ &= \text{AN}\sin\theta + \text{NP}\sin\theta \\ &= (\text{AN} + \text{NP})\sin\theta \\ &= \text{AP}\sin\theta = x_2 \sin\theta. \end{aligned}$$

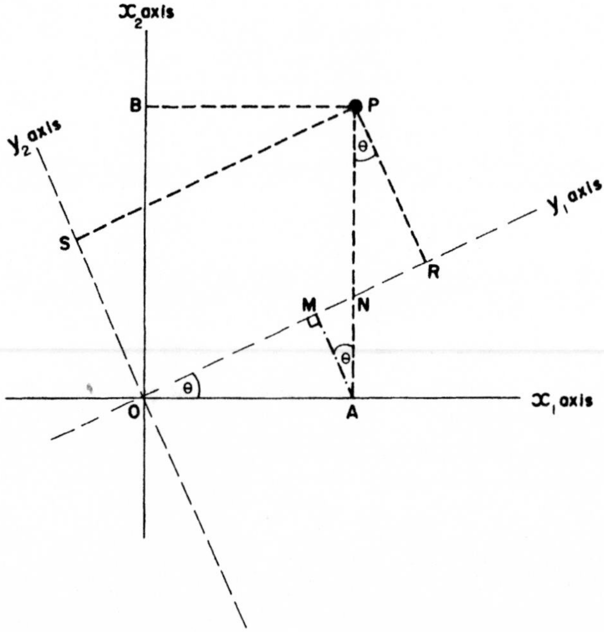

Figure 3.4. Illustrating the conversion of the coordinates of point P relative to the x-axes to coordinates relative to the y-axes.

Therefore,

$$y_1 = x_1 \cos\theta + x_2 \sin\theta. \tag{3.4a}$$

Exactly analogous arguments (which the reader should check) show that

$$y_2 = -x_1 \sin\theta + x_2 \cos\theta \tag{3.4b}$$

Thus the pair of Equations (3.4a) and (3.4b) give the y-coordinates of the point in terms of the x-coordinates and the angle θ. Writing the pair of equations as a single equation representing the equality of two two-element column vectors gives

$$\begin{pmatrix} y_1 \\ y_2 \end{pmatrix} = \begin{pmatrix} x_1 \cos\theta + x_2 \sin\theta \\ -x_1 \sin\theta + x_2 \cos\theta \end{pmatrix}$$

Here the left-hand side is **y**. The right-hand side is the matrix product

$$\begin{pmatrix} \cos\theta & \sin\theta \\ -\sin\theta & \cos\theta \end{pmatrix} \begin{pmatrix} x_1 \\ x_2 \end{pmatrix} = \mathbf{Ux} \tag{3.5}$$

in which the 2×2 matrix **U** has elements

$$u_{11} = u_{22} = \cos\theta; \qquad u_{12} = \sin\theta; \qquad u_{21} = -\sin\theta.$$

We have now discovered how to construct all possible 2×2 orthogonal matrices. All have the form

$$\begin{pmatrix} \cos\theta & \sin\theta \\ -\sin\theta & \cos\theta \end{pmatrix};$$

θ is the angle through which the axes are rotated. In the example on page 94, $\theta = 36.87°$, whence $\cos\theta = 0.8$ and $\sin\theta = 0.6$.

An important property of orthogonal matrices must now be described. First, a definition is needed: the *transpose* of any matrix is the matrix obtained by writing its rows as columns or, equivalently, its columns as rows. For example, the transpose of the 2×3 matrix **A**, where

$$\mathbf{A} = \begin{pmatrix} a_{11} & a_{12} & a_{13} \\ a_{21} & a_{22} & a_{23} \end{pmatrix},$$

is the 3×2 matrix

$$\mathbf{A}' = \begin{pmatrix} a_{11} & a_{21} \\ a_{12} & a_{22} \\ a_{13} & a_{23} \end{pmatrix}.$$

Obviously, the transpose of an $s \times n$ matrix (for instance) is an $n \times s$ matrix. The transpose of a matrix is always denoted by the same symbol as the untransposed matrix with a prime added. Thus the transpose of **A** is denoted by **A**′, and the transpose of a column vector, say **x**, is the row vector **x**′.

Now let us obtain **U**′ the transpose of **U** in Equation (3.5) and then form the product **UU**′.

$$\mathbf{U}' = \begin{pmatrix} \cos\theta & -\sin\theta \\ \sin\theta & \cos\theta \end{pmatrix}$$

and, therefore,

$$\mathbf{UU}' = \begin{pmatrix} \cos\theta & \sin\theta \\ -\sin\theta & \cos\theta \end{pmatrix}\begin{pmatrix} \cos\theta & -\sin\theta \\ \sin\theta & \cos\theta \end{pmatrix}$$

$$= \begin{pmatrix} \cos^2\theta + \sin^2\theta & -\sin\theta\cos\theta + \sin\theta\cos\theta \\ -\sin\theta\cos\theta + \sin\theta\cos\theta & \sin^2\theta + \cos^2\theta \end{pmatrix}$$

$$= \begin{pmatrix} 1 & 0 \\ 0 & 1 \end{pmatrix}$$

or

$$\mathbf{UU}' = \mathbf{I} \tag{3.6}$$

since $\cos^2\theta + \sin^2\theta = 1$.

Equation (3.6) is true, in general, of all orthogonal matrices of any size. Orthogonal matrices are always square. Before discussing the general $s \times s$ orthogonal matrix, it is desirable to make a small change in the symbols. The required change is shown in Figure 3.5. It is seen that the angles have been relabeled thus:

θ_{11} is the angle between the y_1-axis and the x_1-axis;
θ_{12} is the angle between the y_1-axis and the x_2-axis;
θ_{21} is the angle between the y_2-axis and the x_1-axis;
θ_{22} is the angle between the y_2-axis and the x_2-axis.

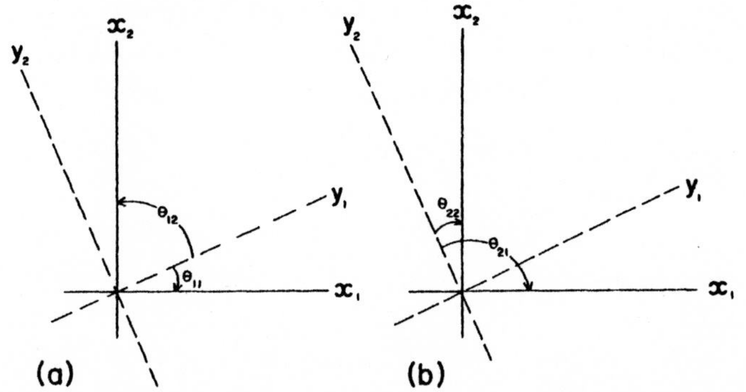

Figure 3.5. The angles between the x-axes and the y-axes: (a) The y_1-axis makes angles θ_{11} and θ_{12} with the x_1- and x_2- axes; (b) The y_2-axis makes angles θ_{21} and θ_{22} with the x_1- and x_2- axes.

It is obvious from the figure that $\theta_{11} = \theta_{22}$ is the same as the original θ; also that

$$\theta_{12} = 90° - \theta \quad \text{or} \quad \theta = 90° - \theta_{12};$$

and

$$\theta_{21} = 90° + \theta \quad \text{or} \quad \theta = \theta_{21} - 90°.$$

The reason for giving every angle a separate symbol becomes clear when we discuss the s-dimensional case.

Consider, now, how the change in symbols affects **U**. The old and new versions are as follows:

$$\mathbf{U} = \begin{pmatrix} \cos\theta & \sin\theta \\ -\sin\theta & \cos\theta \end{pmatrix} = \begin{pmatrix} \cos\theta_{11} & \cos\theta_{12} \\ \cos\theta_{21} & \cos\theta_{22} \end{pmatrix}. \tag{3.7}$$

This result is reached using the relationships:

$$u_{11} = \cos\theta = \cos\theta_{11};$$

$$u_{12} = \sin\theta = \sin(90° - \theta_{12}) = \cos\theta_{12};$$

$$u_{21} = -\sin\theta = -\sin(\theta_{21} - 90°) = \cos\theta_{21};$$

$$u_{22} = \cos\theta = \cos\theta_{22}.$$

These equations express each u_{ij} as the cosine of the corresponding angle θ_{ij}.

It is now intuitively clear how an $s \times s$ orthogonal matrix $\mathbf{U}$ should be constructed (the proof is beyond the scope of this book). Thus

$$\underset{(s\times s)}{\mathbf{U}} = \begin{pmatrix} \cos\theta_{11} & \cos\theta_{12} & \dots & \cos\theta_{1s} \\ \cos\theta_{21} & \cos\theta_{22} & \dots & \cos\theta_{2s} \\ \dots & \dots & \dots & \dots \\ \cos\theta_{s1} & \cos\theta_{s2} & \dots & \cos\theta_{ss} \end{pmatrix}.$$

Equation (3.6), namely,

$$\underset{(s\times s)}{\mathbf{U}}\;\underset{(s\times s)}{\mathbf{U}'} = \underset{(s\times s)}{\mathbf{I}}$$

remains true. This equation states the diagnostic property of orthogonal matrices: that is, a square matrix $\mathbf{U}$ is orthogonal if and only if $\mathbf{UU}' = \mathbf{I}$.

When $\mathbf{U}$ is orthogonal, the transformation

$$\underset{(s\times s)}{\mathbf{U}}\;\underset{(s\times n)}{\mathbf{X}} = \underset{(s\times n)}{\mathbf{Y}}$$

brings about a rigid rotation of the s-dimensional coordinate axes on which are measured the coordinates of an s-dimensional swarm of n data points; their coordinates are given by the columns of $\mathbf{X}$. The coordinates of the points relative to the new coordinate axes, which are still s-dimensional, are given by the columns of $\mathbf{Y}$.

Finally, it should be remarked that the elements of $\mathbf{U}$ are known as the *direction cosines* of the new axes relative to the old. For instance, the element u_{ij} of $\mathbf{U}$, which is $u_{ij} = \cos\theta_{ij}$, is the direction cosine of the y_i-axis relative to the x_j-axis in s-space.

The problem of finding numerical values for the elements of an orthogonal matrix when $s > 2$ is dealt with in Section 3.4. One cannot, as in the two-dimensional case described in detail previously, simply choose one angle and derive all the elements of $\mathbf{U}$ from it. A rotation in s-space requires that s angles be known, and because they are mutually dependent, they cannot be chosen arbitrarily.

3.3. THE PRODUCT OF A DATA MATRIX AND ITS TRANSPOSE

In this section we consider matrices of the form $\mathbf{XX}'$ and $\mathbf{X}'\mathbf{X}$. These are the matrix products formed when a data matrix $\mathbf{X}$ is postmultiplied and premultiplied, respectively, by the transpose of itself. If $\mathbf{X}$ is an $s \times n$

matrix, then $\mathbf{XX}'$ is an $s \times s$ matrix and $\mathbf{X}'\mathbf{X}$ an $n \times n$ matrix. These matrices are needed in many ordination procedures as explained in Chapter 4. For clarity we consider $\mathbf{XX}'$ in detail first, and note the analogous properties of $\mathbf{X}'\mathbf{X}$ subsequently.

The Variance–Covariance Matrix

As always, we denote a data matrix by the symbol **X**. Its elements are in "raw" form; that is, they are the data as recorded in the field. The (i, j)th element is the quantity of species i in quadrat j; i ranges from 1 to s and j ranges from 1 to n.

We now require a *centered* data matrix $\mathbf{X_R}$.* Its (i, j)th element is the amount by which species i in quadrat j deviates from the mean quantity of species i in all n quadrats. Thus the (i, j)th element is $x_{ij} - \bar{x}_i$ where $\bar{x}_i = (1/n)\sum_{j=1}^{n} x_{ij}$. That is, $\bar{x}_i$ is the mean quantity of species i averaged over the n quadrats; equivalently, it is the mean of the n elements in the ith row of **X**. A simple example in which **X** is a 3×4 matrix (Data Matrix #8) is shown in the upper panel of Table 3.2. Because of the way in which it is constructed, all rows of $\mathbf{X_R}$ must sum to zero.

Now form the product

$$\mathbf{R} = \mathbf{X_R X_R'}$$

where $\mathbf{X_R'}$ is the transpose of $\mathbf{X_R}$. **R** is a square $s \times s$ matrix; the ith element on its main diagonal (i.e., its (i, i)th element) is obtained, as usual, by postmultiplying the s-element row vector constituting the ith row of $\mathbf{X_R}$ by the s-element column vector constituting the ith column of $\mathbf{X_R'}$. Since $\mathbf{X_R'}$ is the transpose of $\mathbf{X_R}$, these vectors are "the same" except that one is a row vector and the other a column vector. Thus their product is

$$(x_{i1} - \bar{x}_i, x_{i2} - \bar{x}_i, \ldots, x_{in} - \bar{x}_i)\begin{pmatrix} x_{i1} - \bar{x}_i \\ x_{i2} - \bar{x}_i \\ \vdots \\ x_{in} - \bar{x}_i \end{pmatrix} = \sum_{j=1}^{n} (x_{ij} - \bar{x}_i)^2 \quad (3.8a)$$

*The symbol **R** is used as subscript in $\mathbf{X_R}$ and $\mathbf{S_R}$, and also by itself, because the procedures described are part of an R-type analysis.

TABLE 3.2. DATA MATRIX #8, IN RAW FORM X, AND CENTERED $X_{(c)}$.

$$\bar{x}_1 = \frac{1}{4}\sum_j x_{1j} = 9$$

$$\mathbf{X} = \begin{pmatrix} 4 & 8 & 10 & 14 \\ 17 & 11 & 3 & 1 \\ 2 & 5 & 5 & 4 \end{pmatrix} \qquad \bar{x}_2 = \frac{1}{4}\sum_j x_{2j} = 8$$

$$\bar{x}_3 = \frac{1}{4}\sum_j x_{3j} = 4$$

$$\mathbf{X_R} = \begin{pmatrix} -5 & -1 & 1 & 5 \\ 9 & 3 & -5 & -7 \\ -2 & 1 & 1 & 0 \end{pmatrix}$$

The SSCP matrix **R** and the covariance matrix $(1/n)\mathbf{R}$.

$$\mathbf{R} = \mathbf{X_R X_R'} = \begin{pmatrix} -5 & -1 & 1 & 5 \\ 9 & 3 & -5 & -7 \\ -2 & 1 & 1 & 0 \end{pmatrix} \begin{pmatrix} -5 & 9 & -2 \\ -1 & 3 & 1 \\ 1 & -5 & 1 \\ 5 & -7 & 0 \end{pmatrix}$$

$$= \begin{pmatrix} 52 & -88 & 10 \\ -88 & 164 & -20 \\ 10 & -20 & 6 \end{pmatrix}$$

$$\frac{1}{n}\mathbf{R} = \begin{pmatrix} \text{var}(x_1) & \text{cov}(x_1, x_2) & \text{cov}(x_1, x_3) \\ \text{cov}(x_2, x_1) & \text{var}(x_2) & \text{cov}(x_2, x_3) \\ \text{cov}(x_3, x_1) & \text{cov}(x_3, x_2) & \text{var}(x_3) \end{pmatrix} = \begin{pmatrix} 13 & -22 & 2.5 \\ -22 & 41 & -5 \\ 2.5 & -5 & 1.5 \end{pmatrix}$$

Notice that if the right-hand side were divided by n,* it would give the *variance* of the observations $x_{i1}, x_{i2}, \ldots, x_{in}$, that is, the variance of the variable "quantity of species i per quadrat." It should be recalled that the variance of a variable is the average of the squared deviations of the observations from their mean. In symbols, the variance of the quantity of species i per quadrat is

$$\text{var}(x_i) = \frac{1}{n}\sum_{j=1}^{n}(x_{ij} - \bar{x}_i)^2.$$

The *standard deviation* of this variable, say σ_i, is the square root of the

*n is used as divisor since we are assuming that the n quadrats examined constitute the total population of interest. If the quadrats are a sample of size n from a larger "parent population" for which the variance is to be estimated, the divisor would be $n - 1$.

variance, or

$$\text{var}(x_i) = \sigma_i^2.$$

Next consider the (h, i)th element of **R**. The switch from the familiar symbol pair (i, j) to the unfamiliar (h, i) is because h and i both represent species, namely, the hth and ith species, whereas in the pair (i, j) as used hitherto in this chapter, i denotes a species and j a quadrat. The (h, i)th element of **R** is

$$(x_{h1} - \bar{x}_h, x_{h2} - \bar{x}_h, \ldots, x_{hn} - \bar{x}_h)\begin{pmatrix} x_{i1} - \bar{x}_i \\ x_{i2} - \bar{x}_i \\ \vdots \\ x_{in} - \bar{x}_i \end{pmatrix}$$

$$= \sum_{j=1}^{n} (x_{hj} - \bar{x}_h)(x_{ij} - \bar{x}_i) \tag{3.8b}$$

This is n times the *covariance* of species h and species i in the n quadrats, written $\text{cov}(x_h, x_i)$. When two variables (such as species quantities) are observed on each of n sampling units (such as quadrats), the covariance of the variables is the mean of n cross-products. The cross-product for species h and i in quadrat j is the product of the deviation from its mean of the amount of species h in quadrat j (which is $x_{hj} - \bar{x}_h$) and the deviation from its mean of the amount of species i in quadrat j (which is $x_{ij} - \bar{x}_i$). For given h and i, there are n such cross-products, one for every quadrat, and their average* is the covariance $\text{cov}(x_h, x_i)$. There are $s(s - 1)/2$ covariances altogether, one for every pair of species. Notice that if we put $h = i$ and calculate $\text{cov}(x_i, x_i)$, it is identical with $\text{var}(x_i)$.

Writing **R** out in full, it is seen that

$$\mathbf{R} = \begin{pmatrix} \sum(x_{1j} - \bar{x}_1)^2 & \cdots & \sum(x_{1j} - \bar{x}_1)(x_{sj} - \bar{x}_s) \\ \cdots & \cdots & \cdots \\ \sum(x_{sj} - \bar{x}_s)(x_{1j} - \bar{x}_1) & \cdots & \sum(x_{sj} - \bar{x}_s)^2 \end{pmatrix}$$

*As with the variance, the divisor used to calculate this average is n when the n quadrats are treated as a whole "population," and $n - 1$ when the quadrats are treated as a sample from which the covariance in a larger population is to be estimated.

where all the summations are from $j = 1$ to n. **R** is known as a *sum-of-squares-and-cross-products matrix* or an *SSCP matrix* for short. It is an $s \times s$ matrix.

Alternatively, one may write

$$\mathbf{R} = n \begin{pmatrix} \operatorname{var}(x_1) & \operatorname{cov}(x_1, x_2) & \cdots & \operatorname{cov}(x_1, x_s) \\ \operatorname{cov}(x_2, x_1) & \operatorname{var}(x_2) & \cdots & \operatorname{cov}(x_2, x_s) \\ \cdots & \cdots & \cdots & \cdots \\ \operatorname{cov}(x_s, x_1) & \operatorname{cov}(x_s, x_2) & \cdots & \operatorname{var}(x_s) \end{pmatrix} \tag{3.9}$$

Observe that when, as here, a matrix is multiplied by a scalar (in this case the scalar is n), it means that each individual element of the matrix is multiplied by the scalar. Thus the (h, i)th term of **R** is $n \operatorname{cov}(x_h, x_i)$. **R** is a symmetric matrix since, as is obvious from (3.8b),

$$\operatorname{cov}(x_h, x_i) = \operatorname{cov}(x_i, x_h).$$

The matrix $(1/n)\mathbf{R}$ is known as the *variance–covariance matrix* of the data, or often simply as the *covariance matrix*.

The lower panel of Table 3.2 shows the SSCP matrix **R** and the covariance matrix $(1/n)\mathbf{R}$ for the 3×4 data matrix in the upper panel.

To calculate the elements of a covariance matrix, one may use the procedure demonstrated in Table 3.2, or the computationally more convenient procedure described at the end of this section.

The Correlation Matrix

In the raw data matrix **X** previously discussed the elements are the measured quantities of the different species in each of a sample of quadrats or other sampling units. Often, it is either necessary or desirable to *standardize* these data, that is, rescale the measurements to a standard scale.

Standardization is necessary if different species are measured by different methods in noncomparable units. For example, in vegetation sampling it may be convenient to use cover as the measure of quantity for some species, and numbers of individuals for other species; there is no objection to using incommensurate units such as these provided the data are standardized before analysis.

Standardization is sometimes desirable even when the same units (e.g., numbers of individuals) are used for the measurement of all species quantities. It has the effect of weighting the species according to their rarity so that rare species have as big an influence as common ones on the results of an ordination. Sometimes this is desirable, sometimes not. One may or may not wish to prevent the common species from dominating an analysis. It is a matter of ecological judgment. A thorough discussion of the pros and cons of data standardization has been given by Noy-Meir, Walker, and Williams (1975).

The usual way of standardizing, or rescaling, the data is to divide the observed measurements on each species, after they have been centered (transformed to deviations from the respective species means), by the standard deviation of the species quantities. Thus the element x_{ij} in $\mathbf{X}$ is replaced by

$$\frac{x_{ij} - \bar{x}_i}{\sqrt{\mathrm{var}(x_i)}} = \frac{x_{ij} - \bar{x}_i}{\sigma_i} = z_{ij},$$

say.

We now denote the standardized matrix by $\mathbf{Z_R}$, and examine the product $\mathbf{Z_R Z'_R} = \mathbf{S_R}$, say. ($\mathbf{Z'_R}$ is the transpose of $\mathbf{Z_R}$.)

The (h, i)th element of $\mathbf{S_R}$ is the product of the hth row of $\mathbf{Z_R}$ postmultiplied by the ith column of $\mathbf{Z'_R}$. Thus it is

$$\left(\frac{x_{h1} - \bar{x}_h}{\sigma_h}, \frac{x_{h2} - \bar{x}_h}{\sigma_h}, \dots, \frac{x_{hn} - \bar{x}_h}{\sigma_h}\right) \begin{pmatrix} \dfrac{x_{i1} - \bar{x}_i}{\sigma_i} \\ \dfrac{x_{i2} - \bar{x}_i}{\sigma_i} \\ \vdots \\ \dfrac{x_{in} - \bar{x}_i}{\sigma_i} \end{pmatrix}$$

$$= \frac{\sum_{j=1}^{n} (x_{hj} - \bar{x}_h)(x_{ij} - \bar{x}_i)}{\sigma_h \sigma_i}$$

$$= n \frac{\mathrm{cov}(x_h, x_i)}{\sqrt{\mathrm{var}(x_h)\mathrm{var}(x_i)}}$$

$$= n r_{hi}$$

where r_{hi} is the *correlation coefficient* between species h and species i in the n quadrats.

Observe that the (i, i)th element of $\mathbf{S_R}$ is n. This follows from the fact that $\text{cov}(x_i, x_i) = \text{var}(x_i)$.

The *correlation matrix* is obtained by dividing every element of $\mathbf{S_R}$ by n. Thus

$$\frac{1}{n}\mathbf{S_R} = \begin{pmatrix} 1 & r_{12} & \cdots & r_{1s} \\ r_{21} & 1 & \cdots & r_{2s} \\ \cdots & \cdots & \cdots & \cdots \\ r_{s1} & r_{s2} & \cdots & 1 \end{pmatrix}$$

It is a symmetric matrix since, obviously, $r_{hi} = r_{ih}$.

Table 3.3 shows the standardized form $\mathbf{Z_R}$ of Data Matrix #8 (see Table 3.2), its SSCP matrix $\mathbf{Z_R Z'_R}$, and its correlation matrix. The elements of the correlation matrix may be evaluated either by postmultiplying $\mathbf{Z_R}$ by $\mathbf{Z'_R}$ and dividing by n, or else by dividing the (h, i)th element of the covariance

TABLE 3.3. COMPUTATION OF THE CORRELATION MATRIX FOR DATA MATRIX #8 (SEE TABLE 3.2).

The standardized data matrix is

$$\mathbf{Z_R} = \begin{pmatrix} \frac{-5}{\sqrt{13}} & \frac{-1}{\sqrt{13}} & \frac{1}{\sqrt{13}} & \frac{5}{\sqrt{13}} \\ \frac{9}{\sqrt{41}} & \frac{3}{\sqrt{41}} & \frac{-5}{\sqrt{41}} & \frac{-7}{\sqrt{41}} \\ \frac{-2}{\sqrt{1.5}} & \frac{1}{\sqrt{1.5}} & \frac{1}{\sqrt{1.5}} & 0 \end{pmatrix}.$$

The SSCP matrix for the standardized data is

$$\mathbf{S_R} = \mathbf{Z_R Z'_R} = \begin{pmatrix} 4 & -3.8117 & 2.2646 \\ -3.8117 & 4 & -2.5503 \\ 2.2646 & -2.5503 & 4 \end{pmatrix}.$$

The correlation matrix is

$$\frac{1}{n}\mathbf{S_R} = \begin{pmatrix} 1 & r_{12} & r_{13} \\ r_{21} & 1 & r_{23} \\ r_{31} & r_{32} & 1 \end{pmatrix} = \begin{pmatrix} 1 & -0.9529 & 0.5661 \\ -0.9529 & 1 & -0.6376 \\ 0.5661 & -0.6376 & 1 \end{pmatrix}.$$

matrix, which is $\operatorname{cov}(x_h, x_i)$, by $\sqrt{\operatorname{var}(x_h)\operatorname{var}(x_i)}$, taking the values of the variances from the main diagonal of the covariance matrix. Thus in the example, the (1, 2)th element of the correlation matrix is $r_{12} = -22/\sqrt{13 \times 41} = -0.9529$.

Yet another way of obtaining the correlation matrix is to note that (in the numerical example) it is given by the product

$$\begin{pmatrix} \sqrt{13} & 0 & 0 \\ 0 & \sqrt{41} & 0 \\ 0 & 0 & \sqrt{1.5} \end{pmatrix} \begin{pmatrix} 13 & -22 & 2.5 \\ -22 & 41 & -5 \\ 2.5 & -5 & 1.5 \end{pmatrix} \begin{pmatrix} \sqrt{13} & 0 & 0 \\ 0 & \sqrt{41} & 0 \\ 0 & 0 & \sqrt{1.5} \end{pmatrix}.$$

In the general case this is the product $(1/n)\mathbf{BRB}$ where $\mathbf{B}$ is the diagonal matrix whose (i, i)th element is $\sqrt{\operatorname{var}(x_i)} = \sigma_i$. Notice that when three matrices are to be multiplied (e.g., when the product **LMN** is to be found), it makes no difference whether one first obtains **LM** and then postmultiplies it by **N**, or first obtains **MN** and then premultiplies it by **L**. All that matters is that the order of the factors be preserved. The rule can be extended to the evaluating of a matrix product of any number of factors.

The R Matrix and the Q Matrix

Up to this point we have been considering the matrix product obtained when a data matrix or its centered or standardized equivalent is *post*multiplied by its transpose. The product, whether it be $\mathbf{XX}'$, $\mathbf{X_R X_R'}$, or $\mathbf{Z_R Z_R'}$, is of size $s \times s$. Now we discuss the $n \times n$ matrix obtained when a data matrix is *pre*multiplied by its transpose.

First, with regard to centering: recall that to form $\mathbf{X_R}$, the matrix $\mathbf{X}$ was centered by row means; that is, it was centered by subtracting, from every element, the mean of its row. Equivalently, the data were centered by species means since each row of $\mathbf{X}$ lists observations on one species.

To center $\mathbf{X}'$, we again center by row means. But this time, centering by row means is equivalent to centering by quadrat means since each row of $\mathbf{X}'$ lists observations on one quadrat. The centered form of $\mathbf{X}'$ will be denoted* by $\mathbf{X_Q'}$. The (j, i)th element of $\mathbf{X_Q'}$ is the amount by which x_{ji}, the quantity in quadrat j of species i, deviates from the average quantity of all s species

*The symbol **Q** is used because the procedures described are part of a Q-type analysis.

TABLE 3.4. DATA MATRIX #8 TRANSPOSED, X′, AND THEN CENTERED BY ROWS, $\mathbf{X}'_Q$.

$$\mathbf{X}' = \begin{pmatrix} 4 & 17 & 2 \\ 8 & 11 & 5 \\ 10 & 3 & 5 \\ 14 & 1 & 4 \end{pmatrix} \qquad \begin{aligned} \bar{x}_1 &= 7.67 \\ \bar{x}_2 &= 8.00 \\ \bar{x}_3 &= 6.00 \\ \bar{x}_4 &= 6.33 \end{aligned}$$

$$\mathbf{X}'_Q = \begin{pmatrix} -3.67 & 9.33 & -5.67 \\ 0 & 3 & -3 \\ 4 & -3 & -1 \\ 7.67 & -5.33 & -2.33 \end{pmatrix}$$

The SSCP matrix **Q** and the covariance matrix $(1/s)\mathbf{Q}$

$$\mathbf{Q} = \mathbf{X}'_Q\mathbf{X}_Q = \begin{pmatrix} 132.67 & 45 & -37 & -64.67 \\ 45 & 18 & -6 & -9 \\ -37 & -6 & 26 & 49 \\ -64.67 & -9 & 49 & 92.67 \end{pmatrix}$$

$$\frac{1}{s}\mathbf{Q} = \begin{pmatrix} 44.22 & 15 & -12.33 & -21.56 \\ 15 & 6 & -2 & -3 \\ -12.33 & -2 & 8.67 & 16.33 \\ -21.56 & -3 & 16.33 & 30.89 \end{pmatrix}$$

in the quadrat. (If a species is absent from a quadrat, it is treated as "present" with quantity zero.)

Table 3.4 (which is analogous to Table 3.2) shows **X′**, the transpose of Data Matrix #8, and its row-centered (quadrat-centered) form $\mathbf{X}'_Q$ in the upper panel. In the lower panel is the SSCP matrix $\mathbf{Q} = \mathbf{X}'_Q\mathbf{X}_Q$ (here $\mathbf{X}_Q$ is the transpose of $\mathbf{X}'_Q$) and the covariance matrix $(1/s)\mathbf{Q}$.

The (j, j)th element of $(1/s)\mathbf{Q}$ is the variance of the species quantities in quadrat j. The (j, k)th element is the covariance of the species quantities in quadrats j and k. These elements are denoted, respectively, by $\text{var}(x_j)$ and $\text{cov}(x_j, x_k)$; the two symbols j and k both refer to quadrats. Notice that $\text{var}(x_j)$ could also be defined as the variance of the elements in the jth row of $\mathbf{X}'_Q$, and $\text{cov}(x_j, x_k)$ as the covariance of the elements in its jth and kth rows.

Next $\mathbf{X}'_Q$ is standardized to give $\mathbf{Z}'_Q$; we then obtain the product $\mathbf{S}_Q = \mathbf{Z}'_Q\mathbf{Z}_Q$ where $\mathbf{Z}_Q$ is the transpose of $\mathbf{Z}'_Q$. Finally, $(1/s)\mathbf{S}_Q$ is the correlation matrix whose elements are the correlations between every pair of quadrats. Table 3.5, which is analogous to Table 3.3, shows the steps in the calculations for Data Matrix #8.

TABLE 3.5. COMPUTATION OF THE CORRELATION MATRIX FOR THE TRANSPOSE OF DATA MATRIX #8.

The standardized form of **X**′ is

$$\mathbf{Z}'_Q = \begin{pmatrix} \frac{-3.67}{\sqrt{44.22}} & \frac{9.33}{\sqrt{44.22}} & \frac{-5.67}{\sqrt{44.22}} \\ 0 & \frac{3}{\sqrt{6}} & \frac{-3}{\sqrt{6}} \\ \frac{4}{\sqrt{8.67}} & \frac{-3}{\sqrt{8.67}} & \frac{-1}{\sqrt{8.67}} \\ \frac{7.67}{\sqrt{30.89}} & \frac{-5.33}{\sqrt{30.89}} & \frac{-2.33}{\sqrt{30.89}} \end{pmatrix}.$$

The SSCP matrix for the standardized data is

$$\mathbf{S}_Q = \mathbf{Z}'_Q\mathbf{Z}_Q = \begin{pmatrix} 3 & 2.7626 & -1.8900 & -1.7497 \\ 2.7626 & 3 & -0.8321 & -0.6611 \\ -1.8900 & -0.8321 & 3 & 2.9948 \\ -1.7497 & -0.6611 & 2.9948 & 3 \end{pmatrix}.$$

The correlation matrix is

$$\frac{1}{s}\mathbf{S}_Q = \begin{pmatrix} 1 & 0.9209 & -0.6300 & -0.5832 \\ 0.9209 & 1 & -0.2774 & -0.2204 \\ -0.6300 & -0.2774 & 1 & 0.9983 \\ -0.5832 & -0.2204 & 0.9983 & 1 \end{pmatrix}.$$

Table 3.6 shows a tabular comparison of the two procedures just discussed. These procedures constitute the basic operations in an R-type and a Q-type analysis. It is for this reason that the respective SSCP matrices have been denoted by **R** and **Q**.

Computation of a Covariance Matrix

This subsection is a short digression on the subject of computations. Readers who are concentrating exclusively on principles and who do not wish to be distracted by practical details should skip to Section 3.4.

Consider **R**, the SSCP matrix used in an R-type analysis. It is the product $\mathbf{X}_R\mathbf{X}'_R$. Instead of evaluating the elements of this product as it stands, it is usually more convenient to note that

$$\mathbf{X}_R\mathbf{X}'_R = \mathbf{XX}' - \overline{\mathbf{X}}\overline{\mathbf{X}}' \tag{3.10}$$

and to evaluate the expression on the right side of this equation.

TABLE 3.6. A COMPARISON BETWEEN PRODUCTS OF THE FORM $\mathbf{XX'}$ AND $\mathbf{X'X}$.[a]

Centered matrix	$\mathbf{X_R}$ is matrix $\mathbf{X}$ centered by rows (species). Its (i, j)th term is $x_{ij} - \bar{x}_i$ where $\bar{x}_i = \frac{1}{n}\sum_{j=1}^{n} x_{ij}$.	$\mathbf{X'_Q}$ is matrix $\mathbf{X'}$ centered by rows (quadrats). Its (j, i)th term (in row j, column i) is $x_{ji} - \bar{x}_j$ where $\bar{x}_j = \frac{1}{s}\sum_{i=1}^{s} x_{ji}$.
SSCP matrix	$\mathbf{R} = \mathbf{X_R X'_R}$ where $\mathbf{X'_R}$ is the transpose of $\mathbf{X_R}$. $\mathbf{R}$ is an $s \times s$ matrix; each of its elements is a sum of n squares or cross-products.	$\mathbf{Q} = \mathbf{X'_Q X_Q}$ where $\mathbf{X_Q}$ is the transpose of $\mathbf{X'_Q}$. $\mathbf{Q}$ is an $n \times n$ matrix; each of its elements is a sum of s squares or cross-products.
Covariance matrix	$\frac{1}{n}\mathbf{R}$ The (i, i)th element $\text{var}(x_i)$ is the variance of the elements in the ith row of $\mathbf{X_R}$ (quantities of species i). The (h, i)th element $\text{cov}(x_h, x_i)$ is the covariance of the hth and ith rows of $\mathbf{X_R}$ (quantities of species h and i).	$\frac{1}{s}\mathbf{Q}$ The (j, j)th element $\text{var}(x_j)$ is the variance of the elements in the jth row of $\mathbf{X'_Q}$ (quantities in quadrat j). The (j, k)th element $\text{cov}(x_j, x_k)$ is the covariance of the jth and kth rows of $\mathbf{X'_Q}$ (quantities in quadrats j and k).

Standardized matrix	$\mathbf{Z_R}$ Its (i, j)th term is $\frac{x_{ij} - \bar{x}_i}{\sqrt{\text{var}(x_i)}} = \frac{x_{ij} - \bar{x}_i}{\sigma_i}$ where σ_i is the standard deviation of the quantities of species i.	$\mathbf{Z'_Q}$ Its (j, i)th term is $\frac{x_{ji} - \bar{x}_j}{\sqrt{\text{var}(x_j)}} = \frac{x_{ji} - \bar{x}_j}{\sigma_j}$ where σ_j is the standard deviation of the quantities in quadrat j.
Correlation matrix	$\frac{1}{n}\mathbf{S_R} = \frac{1}{n}\mathbf{Z_R Z'_R}$ The (h, i)th element of $(1/n)\mathbf{S_R}$ is r_{hi}, the correlation coefficient between the hth and ith rows of $\mathbf{X_R}$ (i.e., between species h and species i). The matrix has 1s on its main diagonal since $r_{hh} = 1$ for all h.	$\frac{1}{s}\mathbf{S_Q} = \frac{1}{s}\mathbf{Z'_Q Z_Q}$ The (j, k)th element of $(1/s)\mathbf{S_Q}$ is r_{jk}, the correlation coefficient between the jth and kth rows of $\mathbf{X'_Q}$ (i.e., between quadrats j and k). The matrix has 1s on its main diagonal since $r_{jj} = 1$ for all j.

[a]Symbols h and i refer to species; symbols j and k refer to quadrats.

Here $\bar{\mathbf{X}}$ is an $s \times n$ matrix in which every element in the ith row is

$$\bar{x}_i = \frac{1}{n} \sum_{j=1}^{n} x_{ij},$$

the mean over all quadrats of species i. Thus $\bar{\mathbf{X}}$ has n identical s-element columns. It is

$$\bar{\mathbf{X}} = \begin{pmatrix} \bar{x}_1 & \bar{x}_1 & \cdots & \bar{x}_1 \\ \bar{x}_2 & \bar{x}_2 & \cdots & \bar{x}_2 \\ \cdots & \cdots & \cdots & \cdots \\ \bar{x}_s & \bar{x}_s & \cdots & \bar{x}_s \end{pmatrix}$$

with n columns.

Hence

$$\underset{(s\times n)(n\times s)}{\bar{\mathbf{X}}\ \bar{\mathbf{X}}'} = \begin{pmatrix} \bar{x}_1 & \bar{x}_1 & \cdots & \bar{x}_1 \\ \bar{x}_2 & \bar{x}_2 & \cdots & \bar{x}_2 \\ \cdots & \cdots & \cdots & \cdots \\ \bar{x}_s & \bar{x}_s & \cdots & \bar{x}_s \end{pmatrix} \begin{pmatrix} \bar{x}_1 & \bar{x}_2 & \cdots & \bar{x}_s \\ \bar{x}_1 & \bar{x}_2 & \cdots & \bar{x}_s \\ \cdots & \cdots & \cdots & \cdots \\ \bar{x}_1 & \bar{x}_2 & \cdots & \bar{x}_s \end{pmatrix}$$

$$= \begin{pmatrix} n\bar{x}_1^2 & n\bar{x}_1\bar{x}_2 & \cdots & n\bar{x}_1\bar{x}_s \\ n\bar{x}_2\bar{x}_1 & n\bar{x}_2^2 & \cdots & n\bar{x}_2\bar{x}_s \\ \cdots & \cdots & \cdots & \cdots \\ n\bar{x}_s\bar{x}_1 & n\bar{x}_s\bar{x}_2 & \cdots & n\bar{x}_s^2 \end{pmatrix}$$

which, like $\mathbf{XX}'$, is an $s \times s$ matrix.

The subtraction in (3.10) is done simply by subtracting every element in $\bar{\mathbf{X}}\bar{\mathbf{X}}'$ from the corresponding element in $\mathbf{XX}'$, that is, the (i, j)th element of the former from the (i, j)th element of the latter.

These computations are illustrated in Table 3.7, in which matrix **R** for Data Matrix #8 is obtained using the right side of Equation (3.10). The table should be compared with Table 3.2 in which **R** was obtained using the left side of (3.10). As may be seen, the results are the same.

Now consider the symbolic representation of this result. The (h, i)th element of $\mathbf{X_R X_R'}$ is [see Equation (3.8b)]

$$\sum_{j=1}^{n} (x_{hj} - \bar{x}_h)(x_{ij} - \bar{x}_i).$$

TABLE 3.7. COMPUTATION OF THE SSCP MATRIX FOR DATA MATRIX #8 USING THE EQUATION $\mathbf{R} = \mathbf{XX}' - \bar{\mathbf{X}}\bar{\mathbf{X}}'$.

$$\mathbf{XX}' = \begin{pmatrix} 4 & 8 & 10 & 14 \\ 17 & 11 & 3 & 1 \\ 2 & 5 & 5 & 4 \end{pmatrix} \begin{pmatrix} 4 & 17 & 2 \\ 8 & 11 & 5 \\ 10 & 3 & 5 \\ 14 & 1 & 4 \end{pmatrix} = \begin{pmatrix} 376 & 200 & 154 \\ 200 & 420 & 108 \\ 154 & 108 & 70 \end{pmatrix}$$

$$\bar{\mathbf{X}}\bar{\mathbf{X}}' = \begin{pmatrix} 9 & 9 & 9 & 9 \\ 8 & 8 & 8 & 8 \\ 4 & 4 & 4 & 4 \end{pmatrix} \begin{pmatrix} 9 & 8 & 4 \\ 9 & 8 & 4 \\ 9 & 8 & 4 \\ 9 & 8 & 4 \end{pmatrix} = \begin{pmatrix} 324 & 288 & 144 \\ 288 & 256 & 128 \\ 144 & 128 & 64 \end{pmatrix}$$

$$\mathbf{R} = \mathbf{XX}' - \bar{\mathbf{X}}\bar{\mathbf{X}}' = \begin{pmatrix} 52 & -88 & 10 \\ -88 & 164 & -20 \\ 10 & -20 & 6 \end{pmatrix}$$

The (h, i)th element of $\mathbf{XX}' - \bar{\mathbf{X}}\bar{\mathbf{X}}'$ is

$$\sum_{j=1}^{n} x_{hj}x_{ij} - n\bar{x}_h\bar{x}_i.$$

We now show that these two expressions are identical. In what follows, all sums are from $j = 1$ to n.

Multiplying the factors in brackets in the first expression shows that

$$\sum(x_{hj} - \bar{x}_h)(x_{ij} - \bar{x}_i) = \sum(x_{hj}x_{ij} - \bar{x}_h x_{ij} - \bar{x}_i x_{hj} + \bar{x}_h\bar{x}_i)$$

$$= \sum x_{hj}x_{ij} - \sum \bar{x}_h x_{ij} - \sum \bar{x}_i x_{hj} + \sum \bar{x}_h \bar{x}_j.$$

Now note that $\Sigma\bar{x}_h x_{ij} = \bar{x}_h \Sigma x_{ij}$ and $\Sigma\bar{x}_i x_{hj} = \bar{x}_i \Sigma x_{hj}$ since $\bar{x}_h$ and $\bar{x}_i$ are constant with respect to j (i.e., are the same for all values of j). Similarly, $\Sigma\bar{x}_h\bar{x}_i = n\bar{x}_h\bar{x}_i$ since it is the sum of n constant terms $\bar{x}_h\bar{x}_i$. Thus

$$\sum(x_{hj} - \bar{x}_h)(x_{ij} - \bar{x}_i) = \sum x_{hj}x_{ij} - \bar{x}_h\sum x_{ij} - \bar{x}_i\sum x_{hj} + n\bar{x}_h\bar{x}_i.$$

Next make the substitutions

$$\sum x_{hj} = n\bar{x}_h \quad \text{and} \quad \sum x_{ij} = n\bar{x}_i.$$

Then

$$\sum(x_{hj} - \bar{x}_h)(x_{ij} - \bar{x}_i) = \sum x_{hj}x_{ij} - \bar{x}_h(n\bar{x}_i) - \bar{x}_i(n\bar{x}_h) + n\bar{x}_h\bar{x}_i$$

$$= \sum x_{hj}x_{ij} - n\bar{x}_h\bar{x}_i$$

as was to be proved.

3.4. THE EIGENVALUES AND EIGENVECTORS OF A SQUARE SYMMETRIC MATRIX

This section resumes the discussion in Section 3.2, where it was shown how the pattern of a swarm of data points can be changed by a linear transformation. It should be recalled that, for illustration, a 2×4 data matrix was considered. The swarm of four points it represented were the vertices of a square. Premultiplication of the data matrix by various 2×2 matrices brought about changes in the position or shape of the swarm; see Figure 3.2. When the transforming matrix was orthogonal it caused a rigid rotation of the swarm around the origin of the coordinates (page 94); and when the transformation matrix was diagonal it caused a change in the scales of the coordinate axes.

Clearly, one can subject a swarm of data points to a sequence of transformations one after another, as is demonstrated in the following. The relevance of the discussion to ecological ordination procedures will become clear subsequently.

Rotating and Rescaling a Swarm of Data Points

To begin, consider the following sequence of transformations:

1. Premultiplication of a data matrix **X** by an orthogonal matrix **U**. (Throughout the rest of this book, the symbol **U** always denotes an orthogonal matrix.)
2. Premultiplication of **UX** by a diagonal matrix Λ (capital Greek lambda; the reason for using this symbol is explained later).
3. Premultiplication of Λ**UX** by **U**′, the transpose of **U**, giving **U**′Λ**UX**.

Here is an example. As in section 3.2, we use a 2×4 data matrix representing a swarm of four points in two-space. This time let

$$\mathbf{X} = \begin{pmatrix} 1 & 5 & 1 & 5 \\ 1 & 1 & 4 & 4 \end{pmatrix}.$$

Thus the data swarm consists of the vertices of a rectangle (see Figure 3.6*a*).

The first transformation is to cause a clockwise rotation of the rectangle around the origin through an angle of 25°. The orthogonal matrix required to produce this rotation is (see page 101)

$$\mathbf{U} = \begin{pmatrix} \cos 25° & \cos 65° \\ \cos 115° & \cos 25° \end{pmatrix} = \begin{pmatrix} 0.9063 & 0.4226 \\ -0.4226 & 0.9063 \end{pmatrix}.$$

It is found that the transformed data matrix is

$$\mathbf{UX} = \begin{pmatrix} 1.329 & 4.954 & 2.597 & 6.222 \\ 0.484 & -1.207 & 3.203 & 1.512 \end{pmatrix}.$$

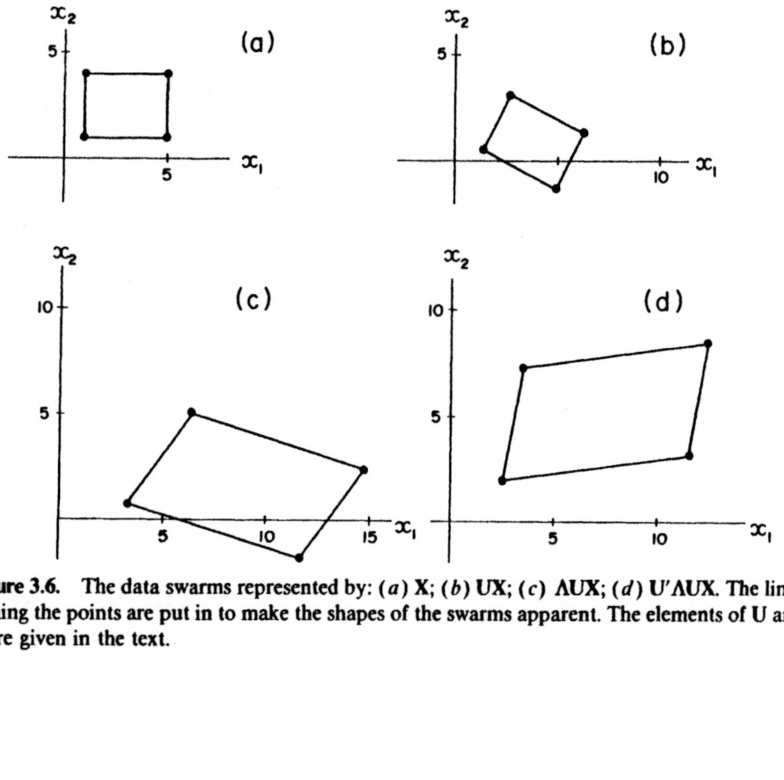

Figure 3.6. The data swarms represented by: (*a*) **X**; (*b*) **UX**; (*c*) **ΛUX**; (*d*) **U′ΛUX**. The lines joining the points are put in to make the shapes of the swarms apparent. The elements of **U** and **Λ** are given in the text.

The swarm of points represented by this matrix, which is merely the original rectangle in a new position, is shown in Figure 3.6*b*.

The second transformation is to be an alteration of the coordinate scales. Let us increase the scale on the x_1-axis (the abscissa) by a factor of $\lambda_1 = 2.4$ and on the x_2-axis (the ordinate) by a factor of $\lambda_2 = 1.6$. This is equivalent to putting

$$\Lambda = \begin{pmatrix} \lambda_1 & 0 \\ 0 & \lambda_2 \end{pmatrix} = \begin{pmatrix} 2.4 & 0 \\ 0 & 1.6 \end{pmatrix}.$$

Then

$$\mathbf{\Lambda UX} = \begin{pmatrix} 3.189 & 11.890 & 6.232 & 14.933 \\ 0.774 & -1.931 & 5.124 & 2.419 \end{pmatrix}.$$

The newly transformed data are plotted in Figure 3.6*c*. The shape of the swarm has changed from a rectangle to a parallelogram.

The third and last transformation consists in rotating the parallelogram back, counterclockwise, through 25°. This is achieved by premultiplying $\mathbf{\Lambda UX}$ by $\mathbf{U}'$, the transpose of $\mathbf{U}$. It is found that

$$\mathbf{U}'\mathbf{\Lambda UX} = \begin{pmatrix} 2.563 & 11.592 & 3.483 & 12.511 \\ 2.049 & 3.275 & 7.278 & 8.504 \end{pmatrix}.$$

These points are plotted in Figure 3.6*d*.

Now observe that we could have achieved the same result by premultiplying the original $\mathbf{X}$ by the matrix $\mathbf{A}$ where

$$\mathbf{A} = \mathbf{U}'\mathbf{\Lambda U} = \begin{pmatrix} 2.2571 & 0.3064 \\ 0.3064 & 1.7429 \end{pmatrix}. \qquad (3.11)$$

Observe that $\mathbf{A}$ is a square symmetric matrix.

We now make the following assertion, without proof. Any $s \times s$ square symmetric matrix $\mathbf{A}$ is the product of three factors that may be written $\mathbf{U}'\mathbf{\Lambda U}$; $\mathbf{U}$ and its transpose $\mathbf{U}'$ are orthogonal matrices; $\mathbf{\Lambda}$ is a diagonal matrix. In the general s-dimensional (or s-species) case, all three factors and $\mathbf{A}$ itself are $s \times s$ matrices.

The elements on the main diagonal of $\mathbf{\Lambda}$ are known as the *eigenvalues* of $\mathbf{A}$ (also called the *latent values* or *roots*, or *characteristic values* or *roots*, of

A). That is, since

$$\Lambda = \begin{pmatrix} \lambda_1 & 0 & \cdots & 0 \\ 0 & \lambda_2 & \cdots & 0 \\ \cdots & \cdots & \cdots & \cdots \\ 0 & 0 & & \lambda_s \end{pmatrix},$$

the eigenvalues of **A** are $\lambda_1, \lambda_2, \ldots, \lambda_s$. The eigenvalues of a matrix are denoted by λs by long-established custom; likewise, the matrix of eigenvalues is always denoted by Λ. This is why Λ was used for the diagonal matrix that rescaled the axes in the second of the three transformations performed previously.

The rows of **U**, which are s-element row vectors, are known as the *eigenvectors* of **A** (also called the *latent vectors*, or *characteristic vectors*, of **A**).

In the preceding numerical example we chose the elements of **U** (hence of **U**′) and Λ, and then obtained **A** by forming the product **U**′Λ**U**. Therefore, we knew, because we had chosen them, the eigenvalues and eigenvectors of this **A** in advance. The eigenvalues are $\lambda_1 = 2.4$ and $\lambda_2 = 1.6$. And the eigenvectors are

$$\mathbf{u}_1' = (0.9063 \quad 0.4226)$$

and

$$\mathbf{u}_2' = (-0.4226 \quad 0.9063),$$

the two rows of **U**.

Now suppose we had started with the square symmetric matrix **A** and had *not* known the elements of **U** and Λ. Would it have been possible to determine **U** and Λ knowing only the elements of **A**? The answer is yes. The analysis which, starting with **A**, finds **U** and Λ such that $\mathbf{A} = \mathbf{U}'\Lambda\mathbf{U}$, in which **U** is orthogonal and Λ diagonal, is known as an *eigenanalysis*. In nearly all ecological ordination procedures, an eigenanalysis forms the heart of the computations, as shown in Chapter 4.

The next step here is a demonstration of one way in which this may be done, first with the 2×2 matrix **A** previously constructed and then with a 3×3 symmetric matrix. The way in which the method can be generalized to permit eigenanalysis of an $s \times s$ symmetric matrix with $s > 3$ will then be clear. Of course, with large s the computations exhaust the patience of anything but a computer.

Hotelling's Method of Eigenanalysis

To begin, recall Equation (3.11)

$$\mathbf{A} = \mathbf{U}'\boldsymbol{\Lambda}\mathbf{U}$$

and premultiply both sides by $\mathbf{U}$ to give

$$\mathbf{UA} = \mathbf{UU}'\boldsymbol{\Lambda}\mathbf{U}.$$

Now observe that, since $\mathbf{U}$ is orthogonal, $\mathbf{UU}' = \mathbf{I}$ by the definition of an orthogonal matrix. Therefore,

$$\mathbf{UA} = \mathbf{I}\boldsymbol{\Lambda}\mathbf{U} = \boldsymbol{\Lambda}\mathbf{U}.$$

Let us write this out in full for the $s = 2$ case. For $\mathbf{U}$ and $\mathbf{A}$ we write each separate element in the customary way, using the corresponding lowercase letter subscripted to show the row and column of the element. For $\boldsymbol{\Lambda}$ we use the knowledge we already possess, namely, that it is a diagonal matrix. Thus

$$\begin{pmatrix} u_{11} & u_{12} \\ u_{21} & u_{22} \end{pmatrix}\begin{pmatrix} a_{11} & a_{12} \\ a_{21} & a_{22} \end{pmatrix} = \begin{pmatrix} \lambda_1 & 0 \\ 0 & \lambda_2 \end{pmatrix}\begin{pmatrix} u_{11} & u_{12} \\ u_{21} & u_{22} \end{pmatrix}.$$

On multiplying out both sides, this becomes

$$\begin{pmatrix} u_{11}a_{11} + u_{12}a_{21} & u_{11}a_{12} + u_{12}a_{22} \\ u_{21}a_{11} + u_{22}a_{21} & u_{21}a_{12} + u_{22}a_{22} \end{pmatrix} = \begin{pmatrix} \lambda_1 u_{11} & \lambda_1 u_{12} \\ \lambda_2 u_{21} & \lambda_2 u_{22} \end{pmatrix}$$

which states the equality of two 2×2 matrices. Not only does the left side (as a whole) equal the right side (as a whole), but it follows also that any row of the matrix on the left side equals the corresponding row of the matrix on the right side. Thus considering only the top row,

$$(u_{11}a_{11} + u_{12}a_{21}, \quad u_{11}a_{12} + u_{12}a_{22}) = (\lambda_1 u_{11}, \quad \lambda_1 u_{12}),$$

an equation having two-element row vectors on both sides. This is the same as the more concise equation

$$\mathbf{u}_1'\mathbf{A} = \lambda_1\mathbf{u}_1'$$

in which $\mathbf{u}_1'$ is the two-element row vector constituting the first row of U, and λ_1 is the only nonzero element in the first row of Λ. Hence λ_1 and $\mathbf{u}_1'$ together are an eigenvalue of **A** and its corresponding eigenvector.

Hotelling's method for obtaining the numerical values of λ_1 and the elements of $\mathbf{u}_1'$ when the elements of **A** are given proceeds as follows. The steps are illustrated using

$$\mathbf{A} = \begin{pmatrix} 2.2571 & 0.3064 \\ 0.3064 & 1.7429 \end{pmatrix},$$

the symmetric matrix whose factors we already know.

Step 1. Choose arbitrary trial values for the elements of $\mathbf{u}_1'$. Denote this trial vector by $\mathbf{w}_{(0)}'$. It is convenient to start with $\mathbf{w}_{(0)}' = (1, 1)$.

Step 2. Form the product $\mathbf{w}_{(0)}'\mathbf{A}$. Thus

$$(1, \quad 1)\mathbf{A} = (2.5635, \quad 2.0493).$$

Let the largest element on the right be denoted by l_1. Thus $l_1 = 2.5635$.

Step 3. Divide each element in the vector on the right by l_1 to give

$$2.5635(1, \quad 0.7994) = l_1\mathbf{w}_{(1)}',$$

say. Now $\mathbf{w}_{(1)}'$ is to be used in place of $\mathbf{w}_{(0)}'$ as the next trial vector.

Step 4. Do steps 2 and 3 again with $\mathbf{w}_{(1)}'$ in place of $\mathbf{w}_{(0)}'$, and obtain l_2 and $\mathbf{w}_{(2)}'$.

Continue the cycle of operations (steps 2 and 3) until a trial vector is obtained that is exactly equal, within a chosen number of decimal places, to the preceding one. Denote this vector by $\mathbf{w}_{(F)}'$ (the subscript F is for "final"). Then the elements of $\mathbf{w}_{(F)}'$ are proportional to the elements of $\mathbf{u}_1'$ the first eigenvector of A, and l_F, the largest element in $\mathbf{w}_{(F)}'\mathbf{A}$, is equal to λ_1, the largest eigenvalue of **A**.

Thus in the example, at the nineteenth cycle and with four decimal places we obtain

$$(1, \quad 0.4663)\mathbf{A} = (2.4000, \quad 1.1191) = 2.4000(1, \quad 0.4663).$$

That is,

$$\mathbf{w}'_{(F)} = (1, \quad 0.4663)$$

and

$$l_F = \lambda_1 = 2.4000.$$

We now wish to obtain $\mathbf{u}'_1$ from $\mathbf{w}'_{(F)}$. Recall (page 102) that $\mathbf{UU}' = \mathbf{I}$, or what comes to the same thing, that the sum of squares of the elements in any row of $\mathbf{U}$ is 1. Hence $\mathbf{u}'_1$ is obtained from $\mathbf{w}'_{(F)}$ by dividing each element in $\mathbf{w}'_{(F)}$ by the square root of the sum of squares of its elements. That is,

$$\mathbf{u}'_1 = \left(\frac{1}{\sqrt{1^2 + 0.4633^2}}, \quad \frac{0.4633}{\sqrt{1^2 + 0.4633^2}} \right)$$

$$= (0.9063, \quad 0.4226).$$

These steps are summarized in Table 3.8.

Having obtained λ_1 and $\mathbf{u}'_1$, the first eigenvalue and eigenvector of $\mathbf{A}$, it is easy (when $s = 2$) to obtain the second pair.

TABLE 3.8. EIGENANALYSIS OF THE 2 × 2 MATRIX A BY HOTELLING'S METHOD.[a]

Cycle Number i	Trial Eigenvector $\mathbf{w}'_{(i)}$	$\mathbf{w}'_{(i)}\mathbf{A} = l_{i+1}\mathbf{w}'_{(i+1)}$
0	(1, 1)	(2.5635, 2.0493) = 2.5635(1, 0.7994)
1	(1, 0.7994)	(2.5020, 1.6997) = 2.5020(1, 0.6793)
2	(1, 0.6793)	(2.4652, 1.4904) = 2.4652(1, 0.6046)
⋮	⋮	⋮
19	(1, 0.4663)	(2.4000, 1.1191) = 2.4000(1, 0.4663)
20	(1, 0.4663)	(2.4000, 1.1191) = 2.4000(1, 0.4663)

Hence $\lambda_1 = 2.4000$
and $\mathbf{u}'_1$ is proportional to (1, 0.4663).

[a]Given $\mathbf{A} = \begin{pmatrix} 2.2571 & 0.3064 \\ 0.3064 & 1.7429 \end{pmatrix}$.

We know (page 101) that $\mathbf{U}$ has the form

$$\mathbf{U} = \begin{pmatrix} \cos\theta & \sin\theta \\ -\sin\theta & \cos\theta \end{pmatrix}$$

and we have just obtained $\mathbf{u}_1'$ the first row of $\mathbf{U}$ which is

$$\mathbf{u}_1' = (0.9063, \quad 0.4226).$$

Therefore,

$$\mathbf{U} = \begin{pmatrix} 0.9063 & 0.4226 \\ -0.4226 & 0.9063 \end{pmatrix}$$

and $\mathbf{u}_2'$ is the second row of $\mathbf{U}$.

To find λ_2, the eigenvalue corresponding to $\mathbf{u}_2'$, recall Equation (3.11), namely,

$$\mathbf{A} = \mathbf{U}'\mathbf{\Lambda}\mathbf{U}.$$

Premultiply both sides by $\mathbf{U}$ and then postmultiply both sides by $\mathbf{U}'$. Hence

$$\mathbf{UAU}' = \mathbf{UU}'\mathbf{\Lambda}\mathbf{UU}'.$$

Since $\mathbf{U}$ is orthogonal, $\mathbf{UU}' = \mathbf{I}$; therefore,

$$\mathbf{UAU}' = \mathbf{I\Lambda I} = \mathbf{\Lambda}. \tag{3.12}$$

Hence we may find λ_2 by evaluating $\mathbf{UAU}'$. It is found that

$$\mathbf{UAU}' = \begin{pmatrix} 2.4 & 0 \\ 0 & 1.6 \end{pmatrix}.$$

Thus $\lambda_1 = 2.4$ (as was found before) and $\lambda_2 = 1.6$.

Next consider the application of Hotelling's method to a 3×3 matrix, say $\mathbf{B}$. Again it is convenient to demonstrate the procedure with a numerical example. Let

$$\mathbf{B} = \begin{pmatrix} 6.0 & 0.2 & 2.4 \\ 0.2 & 5.6 & -0.4 \\ 2.4 & -0.4 & 5.2 \end{pmatrix}.$$

Begin by finding λ_1 and $\mathbf{u}_1'$ as was done with a 2×2 matrix. As before, start with the trial matrix (1, 1, 1). The computations proceed as follows (only 3 decimal places are shown, but 12 were used to obtain these results).

$$(1, \quad 1, \quad 1)\mathbf{B} = (8.6, \quad 5.4, \quad 7.2) = 8.6(1, \quad 0.628, \quad 0.837);$$

$$(1, \quad 0.628, \quad 0.837)\mathbf{B} =$$

$$(8.135, \quad 3.381, \quad 6.502) = 8.135(1, \quad 0.416, \quad 0.799);$$

. . . .

$$(1, \quad -0.058, \quad 0.854)\mathbf{B} =$$

$$(8.038, \quad -0.466, \quad 6.864) = 8.038(1, \quad -0.058, \quad 0.854).$$

Therefore,

$$\lambda_1 = 8.038,$$

and $\mathbf{u}_1'$ is proportional to (1, −0.058, 0.854). Dividing through the elements of this last vector by

$$\sqrt{1^2 + (-0.058)^2 + 0.854^2} = 1.316$$

shows that

$$\mathbf{u}_1' = (0.760, \quad -0.044, \quad 0.649).$$

Now, to find the second eigenvalue and eigenvector, λ_2 and $\mathbf{u}_2'$, proceed as follows. Start by constructing a new matrix $\mathbf{B}_1$; it is known as the first *residual matrix* of $\mathbf{B}$ and is given by

$$\mathbf{B}_1 = \mathbf{B} - \lambda_1 \mathbf{u}_1 \mathbf{u}_1'$$

$$= \mathbf{B} - 8.038 \begin{pmatrix} 0.760 \\ -0.044 \\ 0.649 \end{pmatrix} (0.760, \quad -0.044, \quad 0.649)$$

$$= \begin{pmatrix} 6.0 & 0.2 & 2.4 \\ 0.2 & 5.6 & -0.4 \\ 2.4 & -0.4 & 5.2 \end{pmatrix} - \begin{pmatrix} 4.639 & -0.269 & 3.962 \\ -0.269 & 1.565 & -0.230 \\ 3.962 & -0.230 & 3.383 \end{pmatrix}$$

$$= \begin{pmatrix} 1.361 & 0.469 & -1.562 \\ 0.469 & 4.035 & -0.170 \\ -1.562 & -0.170 & 1.817 \end{pmatrix}.$$

(Note: this is only approximate; for accurate results at the next step, many more decimal places would be needed.)

The values of λ_2 and $\mathbf{u}_2'$ may now be obtained from $\mathbf{B}_1$ in exactly the same way as λ_1 and $\mathbf{u}_1'$ were obtained from $\mathbf{B}$. It is found that

$$\lambda_2 = 5.671 \quad \text{and} \quad \mathbf{u}_2' = (0.144, \quad 0.984, \quad -0.102).$$

Finally, since $\mathbf{B}$ is a 3×3 matrix there is a third eigenvalue–eigenvector pair still to be found, λ_3 and $\mathbf{u}_3'$. To find them, compute $\mathbf{B}_2$ the second residual matrix of $\mathbf{B}$ from the equation

$$\mathbf{B}_2 = \mathbf{B}_1 - \lambda_2 \mathbf{u}_2 \mathbf{u}_2',$$

or, equivalently,

$$\mathbf{B}_2 = \mathbf{B} - \lambda_1 \mathbf{u}_1 \mathbf{u}_1' - \lambda_2 \mathbf{u}_2 \mathbf{u}_2'$$

and operate on $\mathbf{B}_2$ in exactly the same way as $\mathbf{B}$ and $\mathbf{B}_1$ were operated on. It is found that

$$\lambda_3 = 3.092 \quad \text{and} \quad \mathbf{u}_3' = (-0.634, \quad 0.171, \quad 0.754).$$

The eigenanalysis of $\mathbf{B}$ is now complete. As a check, recall that, applying Equation (3.12), we should have

$$\mathbf{UBU}' = \mathbf{\Lambda}.$$

Substituting the numerical results just obtained in the left side of this equation gives

$$\mathbf{UBU}' = \begin{pmatrix} 0.760 & -0.044 & 0.649 \\ 0.144 & 0.984 & -0.102 \\ -0.634 & 0.171 & 0.754 \end{pmatrix} \begin{pmatrix} 6.0 & 0.2 & 2.4 \\ 0.2 & 5.6 & -0.4 \\ 2.4 & -0.4 & 5.2 \end{pmatrix}$$

$$\times \begin{pmatrix} 0.760 & 0.144 & -0.634 \\ -0.044 & 0.984 & 0.171 \\ 0.649 & -0.102 & 0.754 \end{pmatrix}$$

$$= \begin{pmatrix} 8.043 & -0.001 & 0 \\ -0.001 & 5.667 & 0.001 \\ 0 & 0.001 & 3.091 \end{pmatrix} \simeq \begin{pmatrix} 8.038 & 0 & 0 \\ 0 & 5.671 & 0 \\ 0 & 0 & 3.092 \end{pmatrix} = \mathbf{\Lambda}.$$

(The inexactness is because only three decimal places were used.)

It should now be clear how a square symmetric matrix of any size is analyzed by Hotelling's method. The eigenvalues are always obtained in decreasing order of magnitude; that is, given an $s \times s$ matrix, we always* have $\lambda_1 > \lambda_2 > \cdots > \lambda_s$.

A way of reducing the large numbers of computational cycles that are often needed to find each eigenvalue–eigenvector pair is outlined in Tatsuoka (1971); it is beyond the scope of this book.

Finally, it is worth noticing, though it is not proved here, that the sum of the eigenvalues is always equal to the sum of the diagonal terms of the matrix analyzed, which is known as the *trace* of the matrix.

Thus the trace of the 2×2 matrix **A** analyzed before is

$$\text{tr}(\mathbf{A}) = a_{11} + a_{22} = 2.2571 + 1.7429 = 4$$

and

$$\sum_{i=1}^{2} \lambda_i = 2.4 + 1.6 = 4.$$

For the 3×3 matrix **B** we have

$$\text{tr}(\mathbf{B}) = b_{11} + b_{22} + b_{33} = 6.0 + 5.6 + 5.2 = 16.8$$

and

$$\sum_{i=1}^{3} \lambda_i = 8.038 + 5.671 + 3.092 = 16.801$$

(the discrepancy is merely a rounding error).

3.5. THE EIGENANALYSIS OF XX′ AND X′X

It was remarked in Section 3.3 that, in ordinating ecological data, one frequently begins by forming the product **XX′** or **X′X**, where **X** is an $s \times n$ data matrix. Obviously, these two products are related in some way. Each is

*The theoretical possibility of finding a pair of equal eigenvalues is ignored in this elementary discussion.

the product of the same two factors and only the order of the factors differs. Let us put

$$\mathbf{F} = \mathbf{XX'} \quad \text{and} \quad \mathbf{G} = \mathbf{X'X}.$$

(The symbols **R** and **Q** are not used here since **R** and **Q** are the products of row-centered matrices; see Table 3.6. The rows of **X** and **X′** are not centered in forming the products **F** and **G**.)

Clearly, **F** is a symmetric $s \times s$ matrix and **G** is a symmetric $n \times n$ matrix. Suppose we were to do eigenanalyses on both **F** and **G**. How would the results be related?

We first answer this question in symbols and then examine a numerical example. At every step the reader should check that the matrices on each side of an equals sign are of the same size.

Let λ_i be the ith eigenvalue of **F**, and let the s-element row vector $\mathbf{u}'_i$ be the corresponding eigenvector. It follows that

$$\mathbf{u}'_i\mathbf{F} = \lambda_i\mathbf{u}'_i \tag{3.13}$$

or, equivalently,

$$\mathbf{u}'_i\mathbf{XX'} = \lambda_i\mathbf{u}'_i.$$

Postmultiply both sides of this equation by **X** to give

$$\mathbf{u}'_i\mathbf{XX'X} = \lambda_i\mathbf{u}'_i\mathbf{X}.$$

Then, since $\mathbf{X'X} = \mathbf{G}$ by definition, the equation becomes

$$(\mathbf{u}'_i\mathbf{X})\mathbf{G} = \lambda_i(\mathbf{u}'_i\mathbf{X}). \tag{3.14}$$

The factor $\mathbf{u}'_i\mathbf{X}$ on both sides is an n-element row vector.

Comparing (3.13) and (3.14), it is evident that λ_i is an eigenvalue of **G** as well as of **F**, and that the corresponding eigenvector of **G** is either equal to or proportional to $\mathbf{u}'_i\mathbf{X}$.

Thus the eigenvalues and eigenvectors of **G** can be derived from those of **F** or vice versa. This fact is a great aid to computation, especially if either n or s is very large. Thus suppose n greatly exceeds s. A direct eigenanalysis of the $n \times n$ matrix $\mathbf{G} = \mathbf{X'X}$ would entail very long computations if n

were large. The same results could be obtained much faster by analyzing the smaller $s \times s$ matrix $\mathbf{F} = \mathbf{XX}'$.

Now consider a numerical example. The results are given here to 3 decimal places, although 12 were used in the original computations.

The 2×3 data matrix is

$$\mathbf{X} = \begin{pmatrix} 3 & 8 & 7 \\ 11 & 2 & 8 \end{pmatrix}.$$

Then

$$\mathbf{F} = \mathbf{XX}' = \begin{pmatrix} 122 & 105 \\ 105 & 189 \end{pmatrix} \quad \text{and} \quad \mathbf{G} = \mathbf{X}'\mathbf{X} = \begin{pmatrix} 130 & 46 & 109 \\ 46 & 68 & 72 \\ 109 & 72 & 113 \end{pmatrix}.$$

The first eigenvalue and eigenvector of **F**, which can be found by Hotelling's method (or by other methods not described in this book), are

$$\lambda_1 = 265.714 \quad \text{and} \quad \mathbf{u}_1' = (0.590, \quad 0.807).$$

This result can be checked by evaluating both sides of Equation (3.13) and finding that

$$\mathbf{u}_1'\mathbf{F} = \lambda_1\mathbf{u}_1' = (156.755, \quad 214.552).$$

From the previous argument we know that $\lambda_1 = 265.714$ is also an eigenvalue of **G**, and that the corresponding eigenvector is proportional to $\mathbf{u}_1'\mathbf{X}$ which is

$$(0.590, \quad 0.807)\begin{pmatrix} 3 & 8 & 7 \\ 11 & 2 & 8 \end{pmatrix} = (10.652, \quad 6.334, \quad 10.589).$$

Then to find this eigenvector, say $\mathbf{v}_1'$, of **G** it is only necessary to normalize the vector $\mathbf{u}_1'\mathbf{X}$; this is done by dividing every element in $\mathbf{u}_1'\mathbf{X}$ by the square root of the sum of squares of its elements, namely, 16.301. Thus

$$\mathbf{v}_1' = (0.653, \quad 0.389, \quad 0.650),$$

and, as is necessary for an eigenvector, the squares of its elements sum to unity.

As a check that λ_1 and $\mathbf{v}_1'$ are an eigenvalue–eigenvector pair of **G**, note that

$$\mathbf{v}_1'\mathbf{G} = \lambda_1\mathbf{v}_1' = (173.631, \quad 103.255, \quad 172.610).$$

Finding the second eigenvalue of both **F** and **G**, say λ_2, and their respective eigenvectors $\mathbf{u}_2'$ and $\mathbf{v}_2'$ is straightforward. The question now arises as to what is the third eigenvalue λ_3 of the 3×3 matrix **G**, since the 2×2 matrix **F** has no third eigenvalue. The answer is $\lambda_3 = 0$. Further, an eigenanalysis of **G** gives as the third eigenvector $\mathbf{v}_3'$ corresponding to λ_3,

$$\mathbf{v}_3' = (-0.456, \quad -0.483, \quad 0.748)$$

and it will be found that

$$\mathbf{v}_3'\mathbf{G} = \lambda_3\mathbf{G} = (0, \quad 0, \quad 0)$$

(disregarding minor rounding errors).

To summarize: suppose $\mathbf{F} = \mathbf{XX}'$ is an $s \times s$ matrix and $\mathbf{G} = \mathbf{X}'\mathbf{X}$ is an $n \times n$ matrix. Suppose $n > s$ and let $n - s = d$. Then **F** and **G** have s identical eigenvalues and the remaining d eigenvalues of **G** are all zero. The s eigenvectors of **G** that belong with its nonzero eigenvalues can be found from the corresponding eigenvectors of **F**. To do this, note that the ith eigenvector of **G**, say, $\mathbf{v}_i'$, is proportional to $\mathbf{u}_i'\mathbf{X}$, where $\mathbf{u}_i'$ is the ith eigenvector of **F**. The elements of $\mathbf{v}_i'$ are obtained by normalizing the elements of the vector $\mathbf{u}_i'\mathbf{X}$.

EXERCISES

3.1. Consider the following three matrices:

$$\mathbf{A} = \begin{pmatrix} 1 & 2 & 1 \\ 3 & 0 & -1 \end{pmatrix}; \qquad \mathbf{B} = \begin{pmatrix} 4 & 1 & 1 & 2 \\ 3 & -1 & 0 & -1 \\ -2 & 0 & 3 & 3 \end{pmatrix};$$

$$\mathbf{C} = \begin{pmatrix} 0 & 1 \\ 2 & 1 \\ -1 & 2 \\ 5 & 6 \end{pmatrix}.$$

What are the following products? (a) **AB**; (b) **BC**; (c) **AC**; (d) **CA**; (e) **CB**; (f) **BCA**; (g) **CAB**.

3.2. See Figure 3.7. Four data points, A, B, C, and D, have coordinates in two-space given by the data matrix **X**, where

$$\mathbf{X} = \begin{pmatrix} \overset{A}{-1} & \overset{B}{2} & \overset{C}{2} & \overset{D}{-1} \\ 1 & 1 & -1 & -1 \end{pmatrix}.$$

Find the 2 × 2 transformation matrices $\mathbf{U}_1$, $\mathbf{U}_2$, and $\mathbf{U}_3$ that will, respectively, transform **X** into $\mathbf{Y}_1$, $\mathbf{Y}_2$, and $\mathbf{Y}_3$, as shown graphically in separate coordinate frames in Figure 3.7. (The coordinates of all the points are shown on the axes in the figure.)

3.3. Which of the following matrices $\mathbf{U}_1$ and $\mathbf{U}_2$ is orthogonal?

$$\mathbf{U}_1 = \begin{pmatrix} 0.49237 & -0.61546 & 0.61546 \\ 0.84515 & 0.50709 & -0.16903 \\ 0.30985 & -0.51412 & 0.77814 \end{pmatrix}$$

$$\mathbf{U}_2 = \begin{pmatrix} 0.49237 & -0.61546 & 0.61546 \\ 0.84515 & 0.50709 & -0.16903 \\ -0.20806 & 0.60338 & 0.76983 \end{pmatrix}$$

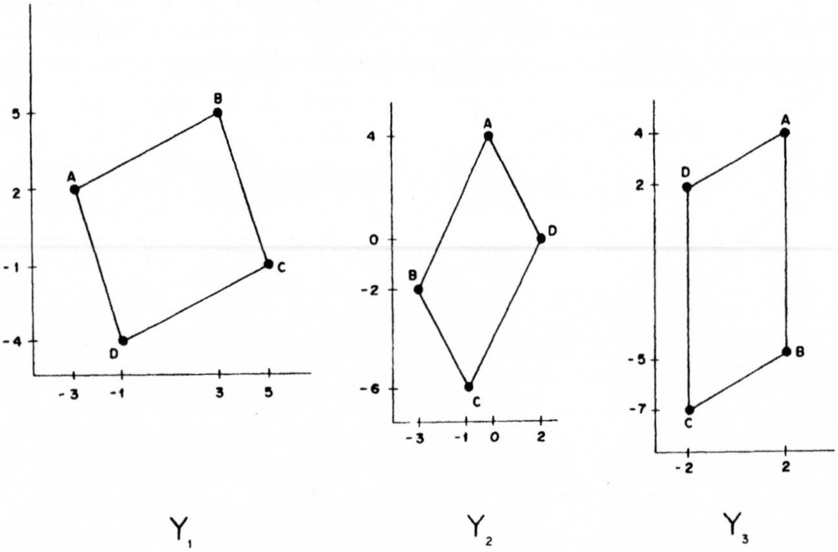

Figure 3.7. See Exercise 3.2.

3.4. Prove that $\mathbf{XX}'$ is symmetric.

3.5. Given the 2×3 data matrix $\mathbf{X}$, where

$$\mathbf{X} = \begin{pmatrix} 2 & 4 & 6 \\ 5 & 1 & 3 \end{pmatrix},$$

find the 2×2 correlation matrix.

3.6. Suppose $\mathbf{A} = \mathbf{U\Lambda U}'$, where $\mathbf{U}$ is orthogonal and the diagonal matrix Λ is

$$\Lambda = \begin{pmatrix} 2 & 0 \\ 0 & 3 \end{pmatrix},$$

whose diagonal elements are the eigenvalues of $\mathbf{A}$. What are the eigenvalues of $\mathbf{A}^5$?

3.7. Eigenanalysis of a 5×5 matrix showed that its first eigenvector was proportional to $(1, 0.87, 0.63, -0.20, -0.11)$. What is this eigenvector in normalized form?

3.8. Eigenanalysis of the following matrix

$$\mathbf{S} = \begin{pmatrix} 10.16 & 6.16 & -7.48 & 0.24 \\ 6.16 & 5.36 & -4.28 & -1.16 \\ -7.48 & -4.28 & 5.84 & -0.12 \\ 0.24 & -1.16 & -0.12 & 1.36 \end{pmatrix}$$

shows that the first, third, and fourth eigenvalues are

$$\lambda_1 = 19.71; \qquad \lambda_3 = 0.44; \qquad \lambda_4 = 0.02.$$

What is λ_2? [Note: there is no need to do an eigenanalysis of $\mathbf{S}$ to answer the question.]

3.9. Show, using symbols, and test with a numerical example of your own devising, that $(\mathbf{XY})' = \mathbf{Y}'\mathbf{X}'$ where $(\mathbf{XY})'$ is the transpose of $\mathbf{XY}$; Show, likewise, that $(\mathbf{ABC})' = \mathbf{C}'\mathbf{B}'\mathbf{A}'$. [Note: these results will be needed in Chapter 4.]

3.10. Consider the numerical example on page 128. The second eigenvector of $\mathbf{F} = \mathbf{XX}'$ is $\lambda_2 = 45.285$. What are the second eigenvalue and eigenvector of $\mathbf{G} = \mathbf{X}'\mathbf{X}$?

Chapter Four

Ordination

4.1. INTRODUCTION

Ordination is a procedure for adapting a multidimensional swarm of data points in such a way that when it is projected onto a two-space (such as a sheet of paper) any intrinsic pattern the swarm may possess becomes apparent. Several different projections onto differently oriented two-spaces may be necessary to reveal all the intrinsic pattern. Projections onto three-spaces to give solid three-dimensional representations of the data can also be made but the results, when reproduced on paper as perspective drawings, are often unclear unless the data points are very few in number.

This definition of ordination may, at first glance, appear to contradict that given at the beginning of Chapter 3. According to the earlier definition, ordination consists in assigning suitably chosen weights to the different species in a many-species community so that a "score" can be calculated for each quadrat (or other sampling unit). Then the quadrats can be ordered ("ordinated") according to their scores, and the result is a one-dimensional ordination.

Often one wants to use two or more different species-weighting systems, and then two or more different one-dimensional ordinations are obtained. These separate results can conveniently be combined as shown in Figure 3.1 in Chapter 3, where two one-dimensional ordinations have been combined to give a two-dimensional ordination in which every point (quadrat) has as its coordinates two scores obtained from the two different weighting systems. Obviously, if one were to use s different weighting systems (where s is the number of species), the result would be an s-dimensional ordination; the swarm of data points would occupy an s-dimensional coordinate frame and by projecting the points onto each axis in turn, one could recreate each of the one-dimensional ordinations yielded by one of the chosen species-weighting systems. More interestingly, one could project it onto one of the two-dimensional planes defined by a pair of axes and obtain a two-dimensional ordination. There are $s(s - 1)/2$ such planes, hence $s(s - 1)/2$ different two-dimensional ordinations would be possible. In practice, probably only a few of them would be interesting.

Now let us consider how species-weighting systems can best be devised. Clearly, if they are chosen arbitrarily and subjectively, there are infinitely many possibilities. What is required is an objective set of rules for assigning weights to the species. A way of arriving at such a set of rules is simply to treat every species in the same way, and conceptually plot the data in s-space; the result is the familiar swarm of n data points in which the coordinates of each point (representing a quadrat) are the amounts it contains of each of the s species. One then treats the swarm as a single entity and adapts it (e.g., by one of the methods described in the following) in a way that seems likely to reveal the intrinsic pattern of the swarm, if it has one, when it is projected onto visualizable two or three-dimensional spaces.

Many methods of adapting a swarm of raw data points have been invented and some of them are described in succeeding sections of this chapter. What unites the methods is that each amounts to a technique for adapting raw observational data in a way that makes them (or is intended to make them) more understandable. Since the initial output of each method is an "adapted" swarm in s dimensions, we again have n data points each with s coordinates; each coordinate of each point is a function of the species quantities in the quadrat represented by that point.

The relationship of the two definitions of ordination should now be clear. When the n columns of an s-row data matrix are plotted as n points in s-space, the pattern of the swarm can be changed by assigning different scores to the species (equivalently, by multiplying each row of the matrix by a different weighting factor). Or, looking at the process the other way round, the swarm as a whole can be modified (to make its internal pattern more clearly perceptible) and, provided this is done appropriately, the effect is to give different weights to the several species.

It is worth noticing the parallel between ordinating a swarm of data points and drawing a two-dimensional geographic map of the whole earth or a large part of it. In both cases the object is to represent a pattern in a space of fewer dimensions that it actually occupies, while at the same time retaining as much as possible of the information it contains and (sometimes) keeping distortion to a minimum. (Distortion is not always a bad thing; see page 190.)

The geographer's problem is, of course, much simpler than the ecologist's since the geographer always starts with a visualizable pattern in only three dimensions, whereas the ecologist starts with an unvisualizable pattern in s dimensions; s is often large, and differs from one case to another. The principle is the same, however. And just as one can choose among a large

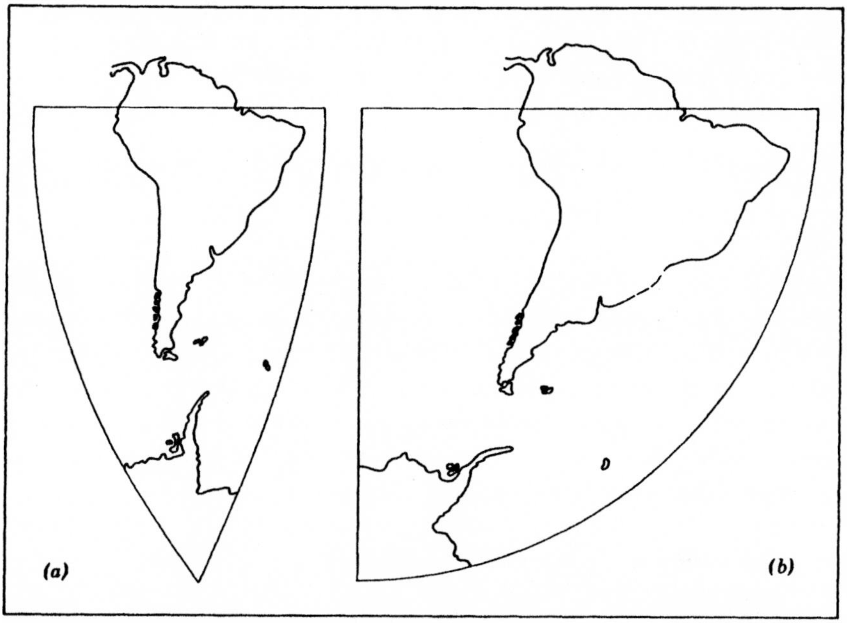

Figure 4.1. Two maps of South America using different map projections illustrating the parallel between constructing a two-dimensional map of part of the three-dimensional globe, and constructing a two-dimensional map (an ordination) of an s-dimensional swarm of data points. The result obtained is very strongly influenced by the technique used. [The map projection for both maps is stereographic (equatorial); the central meridian is at 60°W in (a) and at 120°W in (b).]

number of map projections when drawing a geographic map (see Figure 4.1 for two examples), so one can choose among a large number of ordination techniques when ordinating ecological data. The merits and drawbacks of the various ordination methods have been debated for years and the debate is likely to continue.

The motives for drawing geographic maps are, of course, multifarious: to show climatic data, geological data, bird migration routes, shipping routes, ocean currents, population densities, and so on; the list is endless. But the motive for doing ecological ordinations is always the same, namely, to reveal what is hidden in a body of data, and it is at this point that the parallel between ecology and geography breaks down. A map of South America like that in Figure 4.1*b*, although obviously "wrong" in a way that would require several paragraphs to define precisely, would not mislead a sophisticated map reader. This is because the true shape of South America is thoroughly familiar, and if the distorted two-dimensional version does raise any problems, one can always inspect the ultimate source, a three-dimensional globe. An ecologist does not have these escape hatches: the ecologist's data are always unfamiliar, and (except in rare cases when $s \leq 3$) inspection of the source data is impossible.

4.2. PRINCIPAL COMPONENT ANALYSIS

Principal component analysis (PCA) is the simplest of all ordination methods. The data swarm is projected as it stands, without any differential weighting of the species, onto a differently oriented *s*-space. Equivalently, the axes of the original coordinate frame in which the data points are (conceptually) plotted is rotated rigidly around its origin. This rotation is done in such a way that, relative to the new axes, the pattern of the data swarm shall be, colloquially speaking, as simple as possible.

Ordinating an "Unnatural" Swarm With a Regular Pattern

Before defining the phrase "as simple as possible" exactly, in mathematical terms, it is instructive to continue the discussion at an intuitive level. In the account of PCA that follows we envisage, as a swarm of data points, the eight points at the corners of a cuboid or "box." The fact that such a swarm is utterly unnatural is irrelevant; unnatural assumptions are valuable if they make an argument easier to comprehend. Subsequently (page 142) the procedures devised for analyzing "unnatural" data (the corners of a box) are applied to more believable data swarms, that is, ones that are diffuse and irregular.

Consider the eight points at the corners of the box in Figure 4.2*a* (only seven of the corners are visible in the diagram since, for the sake of clarity, the box is shown as an opaque solid). The center of the box is at the origin of the coordinates. If the points were present alone, without the edges, and were projected onto the plane defined by the x_1- and x_2-axes (the x_1, x_2 plane), there would be a confusing pattern of points with no immediately obvious regularity; the same would be true if they were projected onto the x_1, x_3 plane, or the x_2, x_3 plane. However, if the box were rigidly rotated about the origin of the coordinates until it was oriented as in Figure 4.2*b*, and its corner points were then projected onto the three planes, each

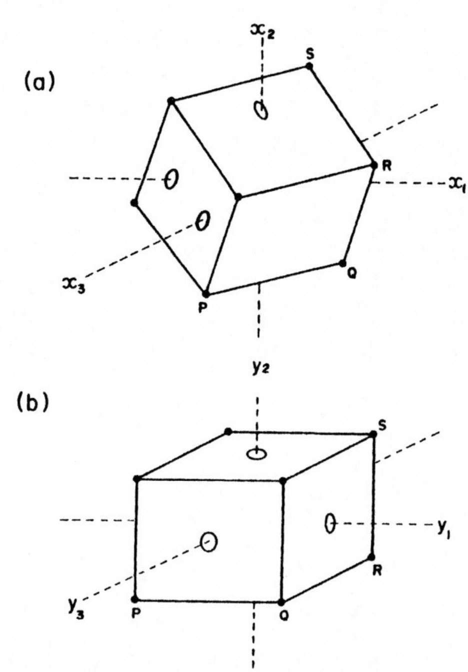

Figure 4.2. A box (cuboid) plotted in a three-dimensional coordinate frame in two different orientations. In (*a*) the box is oblique; in (*b*) it appears after rotation to an orientation that brings its edges parallel with the coordinate axes. The coordinates of the box's corners are denoted by *x*s in the upper graph and by *y*s in the lower graph. The width, height, and depth of the box are PQ, QR, and RS, respectively.

projection would show one of the three faces of the box as a rectangle; the "true" pattern of the points in three-space would be displayed as clearly as possible and the fact that they formed the corners of a box would become obvious.

We now describe the task to be performed in mathematical terms. To repeat, we wish to rotate the box relative to the coordinate frame or, which comes to the same thing, rotate the coordinate frame relative to the box. However, this sentence does not specify in operational terms exactly what needs to be done (unless, of course, one were to do the job physically, with a wood and wires model). The actual operation to be performed consists in finding the coordinates in three-space of the corners of the newly oriented box as shown in Figure 4.2*b* from a knowledge of their original coordinates as in Figure 4.2*a*. We denote the original coordinates by *x*s and the new coordinates after the rotation by *y*s. The axes in Figure 4.2*a* and 4.2*b* are labeled with *x*s and *y*s accordingly.

The original coordinates (three for each of the eight points) form the columns of a 3×8 data matrix. Therefore, to rotate the box, it is necessary to premultiply the data matrix by a 3×3 orthogonal matrix (see Section 3.2) that will bring about the rotation required. The problem, therefore, boils down to finding this orthogonal matrix.

To see how it can be found, notice that the projections of the width, height, and depth of the box (the lengths PQ, QR, and RS, respectively, in Figure 4.2) have their true lengths only when they are projected onto axes parallel with the edges of the box, that is, onto the axes of the coordinate frame when it is oriented as in Figure 4.2*b*. Given this orientation, the projections of the edges on the axes can be seen to be as follows:

$$\begin{array}{l} \text{Projection of edge PQ} \\ \text{(and the other three} \\ \text{edges parallel with it)} \end{array} = \begin{cases} \text{PQ} & \text{on the } y_1\text{-axis} \\ 0 & \text{on the } y_2\text{-axis} \\ 0 & \text{on the } y_3\text{-axis;} \end{cases}$$

$$\begin{array}{l} \text{Projection of edge RS} \\ \text{(and the other three} \\ \text{edges parallel with it)} \end{array} = \begin{cases} 0 & \text{on the } y_1\text{-axis} \\ \text{RS} & \text{on the } y_2\text{-axis} \\ 0 & \text{on the } y_3\text{-axis;} \end{cases}$$

$$\begin{array}{l} \text{Projection of edge QR} \\ \text{(and the other three} \\ \text{edges parallel with it)} \end{array} = \begin{cases} 0 & \text{on the } y_1\text{-axis} \\ 0 & \text{on the } y_2\text{-axis} \\ \text{QR} & \text{on the } y_3\text{-axis.} \end{cases}$$

But given an oblique orientation, like that in Figure 4.2*a* for instance, every edge has a nonzero projection on all three axes, and all these projections are less than the "true" lengths.

We therefore require a rotation of the coordinate frame that will cause each edge of the box to have a nonzero projection (equal to its true length) on one axis only, and zero projections on the other two axes. This requirement specifies in mathematical terms exactly what the desired rotation is to achieve.

To clarify the next stages of the discussion, consider a numerical example and its graphical representation. The box to be rotated is the oblique box in

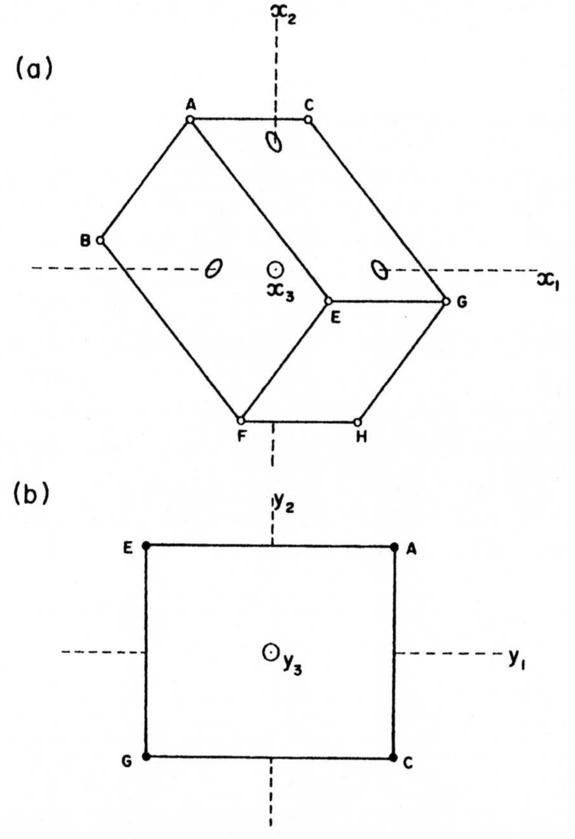

Figure 4.3. The box whose coordinates are given in Table 4.1. (*a*) The box in an oblique position; the coordinates of its corners are the elements of the matrix **X** in the table. (*b*) The box rotated until its edges are parallel with the coordinate axes; the coordinates of its corners are the elements of **Y** in the table. Observe that the x_3- and y_3-axes are perpendicular to the plane of the page. Hence in (*b*) the face ACGE is in the plane of the page; therefore, the width $d(\text{AE}) = 12$ and height $d(\text{AC}) = 10$ of the box are shown without foreshortening.

TABLE 4.1. DATA MATRICES FOR THE POINTS (CORNERS OF A BOX) IN FIGURE 4.3a (MATRIX X) AND FIGURE 4.3b (MATRIX Y).

The data matrix **X**.[a]

$$\mathbf{X} = \begin{pmatrix} \overset{A}{-4.04} & \overset{B}{-8.66} & \overset{C}{1.73} & \overset{D}{-2.88} & \overset{E}{2.88} & \overset{F}{-1.73} & \overset{G}{8.66} & \overset{H}{4.04} \\ 7.07 & 1.41 & 7.07 & 1.41 & -1.41 & -7.07 & -1.41 & -7.07 \\ 3.26 & 0 & -4.89 & -8.16 & 8.16 & 4.89 & 0 & -3.26 \end{pmatrix}$$

The matrix **Y** giving the coordinates of the points as shown in Figure 4.3*b*, after rotation of the box.

$$\mathbf{Y} = \begin{pmatrix} \overset{A}{6} & \overset{B}{6} & \overset{C}{6} & \overset{D}{6} & \overset{E}{-6} & \overset{F}{-6} & \overset{G}{-6} & \overset{H}{-6} \\ 5 & 5 & -5 & -5 & 5 & 5 & -5 & -5 \\ 4 & -4 & 4 & -4 & 4 & -4 & 4 & -4 \end{pmatrix}$$

[a]The capital letter above each column is the label of the corresponding point in Figure 4.3*a*.

Figure 4.3*a*. The three coordinates of its eight corner points are the elements in the columns of the 3×8 data matrix **X** shown in Table 4.1. (The reason for choosing these coordinates becomes clear later.) In Figure 4.3 (in contrast to Figure 4.2) the three-dimensional graphs have their third axes perpendicular to the plane of the page. Therefore, what the drawing in Figure 4.3*a* shows is the projection of the oblique box onto the x_1, x_2 plane.

The size of the box can be found by applying the three-dimensional form of Pythagoras's theorem. Thus $d(\text{AB})$, the length of edge AB which joins the points $\text{A} = (x_{11}, x_{21}, x_{31})$ and $\text{B} = (x_{12}, x_{22}, x_{32})$, is

$$\begin{aligned} d(\text{AB}) &= \sqrt{(x_{11} - x_{12})^2 + (x_{21} - x_{22})^2 + (x_{31} - x_{32})^2} \\ &= \sqrt{(-4.04 + 8.66)^2 + (7.07 - 1.41)^2 + (3.26 - 0)^2} \\ &= 8. \end{aligned}$$

Likewise,

$$\begin{aligned} d(\text{AE}) &= \sqrt{(x_{11} - x_{15})^2 + (x_{21} - x_{25})^2 + (x_{31} - x_{35})^2} \\ &= \sqrt{(-4.04 - 2.88)^2 + (7.07 + 1.41)^2 + (3.26 - 8.66)^2} \\ &= 12. \end{aligned}$$

In the same way, it may be found that $d(\text{AC}) = 10$.

Now suppose that the box is rotated until its edges are parallel with the coordinate axes. Let the rotation bring the box into the position shown in Figure 4.3*b*. The width d(AE) and height d(AC) of the box are unchanged by the rotation and are still 12 and 10 units, respectively; its depth d(AB), which cannot be seen because it is at right angles to the plane of the page, is still 8 units. It is easy to see that the coordinates of the corner points of the newly oriented box are given by the columns of the 3×8 matrix **Y** shown in Table 4.1. Therefore, we now need to find the orthogonal matrix **U** for which

$$\mathbf{UX} = \mathbf{Y}.$$

To do this notice first the form of the product of **Y** postmultiplied by its transpose **Y**′. It is the 3×3 matrix

$$\mathbf{YY}' = \begin{pmatrix} 288 & 0 & 0 \\ 0 & 200 & 0 \\ 0 & 0 & 125 \end{pmatrix},$$

a diagonal matrix. It is diagonal because the points whose coordinates are the columns of **Y** form a box that is aligned with the coordinate axes. (This last point is not proved in this book; it is intuitively reasonable, however, and should seem steadily more reasonable as the rest of this section unfolds.)

Since we are to have **UX** = **Y**, we must also have

$$\mathbf{UX}(\mathbf{UX})' = \mathbf{YY}' \tag{4.1}$$

where $(\mathbf{UX})'$ is the 8×3 transpose of the 3×8 matrix product **UX**. Now recall (from Exercise 3.9) that $(\mathbf{UX})' = \mathbf{X}'\mathbf{U}'$. Thus (4.1) becomes

$$\mathbf{UXX}'\mathbf{U}' = \mathbf{YY}'. \tag{4.2}$$

Next observe that both **XX**′ and **YY**′ are SSCP matrices of the same form as **R** in Section 3.3. Both matrices are, of course, square and symmetric. We use the symbol **R** for **XX**′ and denote **YY**′ by $\mathbf{R_Y}$. Then (4.2) becomes

$$\mathbf{URU}' = \mathbf{R_Y}. \tag{4.3}$$

Finally, compare this equation with Equation (3.12). It is clear that **U**, **U**′,

TABLE 4.2. THE EIGENANALYSIS OF R = XX′.[a]

The SSCP matrix is

$$\mathbf{R} = \mathbf{XX}' = \begin{pmatrix} 205.21 & -65.21 & 3.74 \\ -65.21 & 207.89 & -46.06 \\ 3.74 & -46.06 & 202.25 \end{pmatrix}.$$

The eigenvectors of **R** are the rows of **U** where

$$\mathbf{U} = \begin{pmatrix} -0.57735 & 0.70711 & -0.40825 \\ -0.57735 & 0 & 0.81650 \\ 0.57735 & 0.70711 & 0.40825 \end{pmatrix}.$$

The eigenvalues of **R** are the nonzero elements of $\mathbf{\Lambda}$ where

$$\mathbf{\Lambda} = \mathbf{URU}' = \begin{pmatrix} 288 & 0 & 0 \\ 0 & 200 & 0 \\ 0 & 0 & 128 \end{pmatrix} = \mathbf{R}_Y = \mathbf{YY}'.$$

The coordinates of the box's corners when it is oriented as in Figure 4.3*b* are given by the columns of

$$\mathbf{UX} = \mathbf{Y} = \begin{pmatrix} 6 & 6 & 6 & 6 & -6 & -6 & -6 & -6 \\ 5 & 5 & -5 & -5 & 5 & 5 & -5 & -5 \\ 4 & -4 & 4 & -4 & 4 & -4 & 4 & -4 \end{pmatrix}$$

[a]**X** is the data matrix defining the corner points of the box in Figure 4.3*a*.

and $\mathbf{R}_Y$ can be obtained by doing an eigenanalysis of **R**; the nonzero elements of $\mathbf{R}_Y$ which, as we have seen, are all on the main diagonal are the eigenvalues of **R**.

Table 4.2 shows the outcome of doing an eigenanalysis of **R**. As may be seen, the eigenanalysis gives the eigenvectors of **R**; these are the rows of **U**. **U** is the orthogonal matrix we require. The product **UX** gives **Y**, the matrix of coordinates of the corner points of the box in Figure 4.3*b*, which has been rotated to the desired orientation with its edges parallel with the coordinate axes.

Thus the original problem is solved. To summarize: the solution is found by doing an eigenanalysis of the SSCP matrix **R** = **XX′** where **X** is the original data matrix.

Ordinating a "Natural", Irregular Swarm

Now consider the application of this procedure to "realistic" data swarms. Suppose one had an $s \times n$ data matrix **X** listing the amounts of s species in

n quadrats (or other sampling units) and that these data are thought of as being represented by a swarm of n points in s-space; the swarm is diffuse and irregular. The procedure for performing a PCA on such data is as follows (the symbols used are the same, and have the same meaning, as those in Section 3.3 and Table 3.6. A numerical example using a 2×11 data matrix is shown in Table 4.3).

1. Center the data by species (rows). Do this by subtracting from every element in $\mathbf{X}$ the mean of the elements in the same row. Call the centered data matrix $\mathbf{X_R}$.
2. Form the $s \times s$ SSCP matrix $\mathbf{R} = \mathbf{X_R X_R'}$.
3. Form the $s \times s$ covariance matrix $(1/n)\mathbf{R}$. As we shall see, this step is not strictly necessary, but it is usually done.
4. Carry out an eigenanalysis of $\mathbf{R}$ or $(1/n)\mathbf{R}$. The eigenvectors of these two matrices are identical. Combine these s eigenvectors, each with s elements, by letting them be the rows of an $s \times s$ matrix $\mathbf{U}$; $\mathbf{U}$ is orthogonal. The eigenvalues of $\mathbf{R}$ are n times those of $(1/n)\mathbf{R}$; hence it is immaterial whether $\mathbf{R}$ or $(1/n)\mathbf{R}$ is analyzed. Let Λ denote the $s \times s$ diagonal matrix whose nonzero elements are the eigenvalues of the covariance matrix $(1/n)\mathbf{R}$. Then [compare Equation (3.12)]

$$\mathbf{U}\left(\frac{1}{n}\mathbf{R}\right)\mathbf{U}' = \Lambda.$$

It follows that

$$\mathbf{URU}' = n\Lambda$$

and the nonzero elements of $n\Lambda$ are the eigenvalues of $\mathbf{R}$.

5a. Complete the PCA by forming the $s \times n$ matrix $\mathbf{Y} = \mathbf{UX_R}$. Each column of $\mathbf{Y}$ gives the new set of s coordinates of one of the data points. If the points are plotted using these new coordinates, it is found that the pattern of the points relative to one another is unchanged. The only change produced is that the whole swarm as a single entity has been rotated around its centroid, which is the origin of the new coordinate frame.

Figure 4.4 shows graphically the results in Table 4.3. The original swarm of points whose coordinates are the elements of $\mathbf{X}$ is plotted in Figure 4.4*a*; the transformed swarm, which after PCA has as coordinates the elements of $\mathbf{Y}$, is plotted in Figure 4.4*b*. As may be seen, the centroid of the whole

TABLE 4.3. THE STEPS IN A PRINCIPAL COMPONENTS ANALYSIS OF DATA MATRIX #9.

The 2×11 data matrix is

$$\mathbf{X} = \begin{pmatrix} 20 & 26 & 27 & 28 & 31 & 33 & 39 & 41 & 42 & 48 & 50 \\ 50 & 45 & 60 & 50 & 46 & 55 & 35 & 25 & 33 & 24 & 28 \end{pmatrix}.$$

The row-centered data matrix obtained by subtracting $\bar{x}_1 = 35$ and $\bar{x}_2 = 41$ from the first and second rows of **X**, respectively, is

$$\mathbf{X_R} = \begin{pmatrix} -15 & -9 & -8 & -7 & -4 & -2 & 4 & 6 & 7 & 13 & 15 \\ 9 & 4 & 19 & 9 & 5 & 14 & -6 & -16 & -8 & -17 & -13 \end{pmatrix}.$$

The SSCP matrix is

$$\mathbf{R} = \begin{pmatrix} 934 & -1026 \\ -1026 & 1574 \end{pmatrix}.$$

The covariance matrix is

$$\frac{1}{n}\mathbf{R} = \begin{pmatrix} 84.9091 & -93.2727 \\ -93.2727 & 143.0909 \end{pmatrix}.$$

The matrix of eigenvectors is

$$\mathbf{U} = \begin{pmatrix} 0.592560 & -0.805526 \\ 0.805526 & 0.592560 \end{pmatrix}.$$

The eigenvalues of the covariance matrix are the nonzero elements of

$$\mathbf{\Lambda} = \begin{pmatrix} 211.704 & 0 \\ 0 & 16.295 \end{pmatrix}.$$

The transformed data matrix (after rounding to one decimal place) is

$$\mathbf{Y} = \begin{pmatrix} -16.1 & -8.6 & -20.0 & -11.4 & -6.4 & -12.5 & 7.2 & 16.4 & 10.6 & 21.4 & 19.4 \\ -6.7 & -4.9 & 4.8 & -0.3 & -0.3 & 6.7 & -0.3 & -4.6 & 0.9 & 0.4 & 4.4 \end{pmatrix}$$

swarm (shown by a cross), which is at $(\bar{x}_1, \bar{x}_2) = (35, 41)$ in Figure 4.4*a*, has been shifted to the origin at $(y_1, y_2) = (0, 0)$ in Figure 4.4*b*, and the swarm as a whole has been rotated so that its "long axis" is parallel with the y_1-axis (this statement is expressed more precisely later).

Paragraph 5a describes PCA as a process of rotating a swarm of points around its centroid. It is instructive to rephrase the paragraph, calling it 5b, so that it describes the process as one of rotating the coordinate frame relative to the swarm.

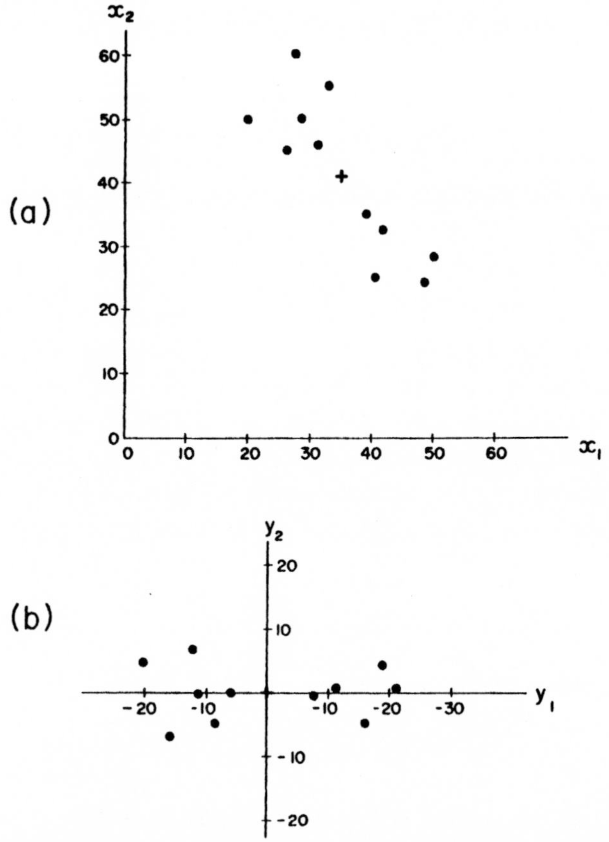

Figure 4.4. Two plots of Data Matrix #9. (*a*) The original untransformed data whose coordinates are given by **X** in Table 4.3. The cross marks the centroid of the swarm. (*b*) After PCA. The origin of the new coordinates is at the swarm's centroid. The swarm has been rotated. The coordinates of each point, measured along axes y_1 and y_2, are given by **Y** in Table 4.3.

5b. Form the matrix product $\mathbf{Y} = \mathbf{U}\mathbf{X_R}$ as already directed. Then plot the original data swarm after centering the raw data by rows; the coordinates are the elements of $\mathbf{X_R}$ (see Figure 4.5). We now wish to rotate the axes rigidly around the origin. This is equivalent to drawing new axes (the y_1 and y_2-axes) which must go through the origin and be perpendicular to each other in such a way that, relative to them, the points shall have coordinates given by $\mathbf{Y}$. The problem, therefore, is to find the equations of the lines in the x_1, x_2 coordinate frame that serve as these new axes.

This is easily done. Note that any imaginable point on the y_2-axis has a coordinate of zero on the y_1-axis, and vice versa. Hence the y_1-axis is the set of all imaginable points, such as point k, of which it is true that

$$y_{2k} = u_{21}x_{1k} + u_{22}x_{2k} = 0.$$

Indeed, the set of all points conforming to this equation *is* the y_1-axis, and its equation is

$$u_{21}x_1 + u_{22}x_2 = 0 \quad \text{or} \quad x_2 = \frac{-u_{21}}{u_{22}}x_1.$$

To draw this straight line, it is obviously necessary to find only two points on it. One point is the origin, at $(x_1, x_2) = (0, 0)$. Another point can be found by assigning any convenient value to x_1 and solving for x_2. In the numerical example we are here considering, for instance, let us put $x_1 = 10$. Then

$$x_2 = \frac{-0.805526}{0.592560} \times 10 = -13.59.$$

Hence two points that define the y_1-axis are

$$(x_1, x_2) = (0, 0) \quad \text{and} \quad (x_1, x_2) = (10, -13.59).$$

This line is shown dashed in Figure 4.5 and is labeled y_1. The y_2-axis is found in the same way. It is the line

$$u_{11}x_1 + u_{12}x_2 = 0 \quad \text{or} \quad x_2 = \frac{-u_{11}}{u_{12}}x_1.$$

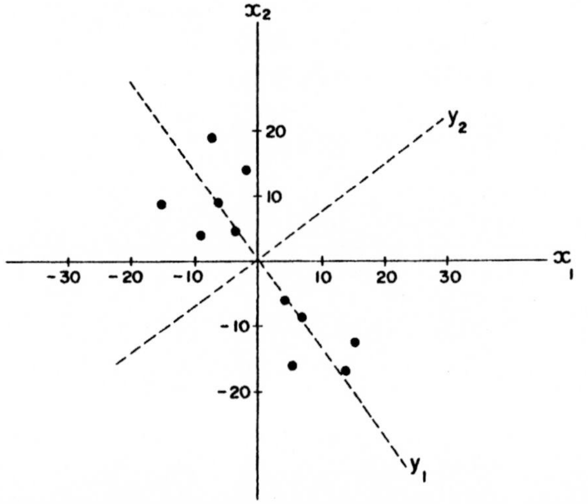

Figure 4.5. Another way of portraying the PCA of Data Matrix #9. The points were plotted using the coordinates in the centered data matrix $\mathbf{X_R}$ (see Table 4.3) with the axes labeled x_1 and x_2. The new axes, the y_1- and y_2-axes, were found as explained in the text. Observe that the pattern of the points relative to the new axes is the same as the pattern in Figure 4.4*b*.

We have now done two PCAs, the first of the corner points of a three-dimensional box, and the second of an irregular two-dimensional swarm of points. It should now be clear that, in general (i.e., for any value of s), doing a PCA consists in doing an eigenanalysis of the SSCP matrix

$$\mathbf{R} = \mathbf{X_R X'_R}$$

or of the covariance matrix

$$\frac{1}{n}\mathbf{R} = \frac{1}{n}\mathbf{X_R X'_R}$$

where $\mathbf{X_R}$ is the row-centered version of the original data matrix.

Then one can either (1) find new coordinates for the n data points from the equation

$$\mathbf{Y} = \mathbf{U X_R},$$

where $\mathbf{U}$ is the orthogonal matrix whose rows are the eigenvectors of $\mathbf{R}$ and $(1/n)\mathbf{R}$; or (2) find the equations of the rotated coordinate axes from the equation

$$\mathbf{Ux} = \mathbf{0},$$

where $\mathbf{x}$ and $\mathbf{0}$ are the s-element column vectors

$$\mathbf{x} = \begin{pmatrix} x_1 \\ x_2 \\ \vdots \\ x_s \end{pmatrix} \quad \text{and} \quad \mathbf{0} = \begin{pmatrix} 0 \\ 0 \\ \vdots \\ 0 \end{pmatrix},$$

and the equation $\mathbf{Ux} = \mathbf{0}$ denotes s equations of which the ith is

$$u_{i1}x_1 + u_{i2}x_2 + \cdots + u_{is}x_s = 0.$$

The ith axis is defined by the $s - 1$ simultaneous equations $\mathbf{u}'_k\mathbf{x} = 0$ with $k = 1, 2, \ldots, (i-1), (i+1), \ldots, s$.

Figures 4.4 and 4.5 amount to graphical demonstrations of alternatives (1) and (2) when $s = 2$.

Regardless of how large s is, there are s mutually perpendicular axes, both for the original coordinate frame and the rotated frame. The new axes, namely, the y_1-axis, the y_2-axis, ..., and the y_s-axis, are known as the first, second, ..., and sth *principal axes* of the data. The new coordinates of the data points measured along these new axes are known as *principal component scores*. For example, the ith principal component score of the jth point is

$$y_{ij} = u_{i1}x_{1j} + u_{i2}x_{2j} + \cdots + u_{is}x_{sj}.$$

Thus it is the weighted sum of the quantities (after they have been centered by species means) of the s species in the jth quadrat. The us are the weights. After PCA each point has as coordinates not the amount of each species in a quadrat, but variously weighted sums of all the species in the quadrat.

The term *principal component* denotes the variable "the principal component score for any data point"; hence the ith principal component of the data is

$$y_i = u_{i1}x_1 + u_{i2}x_2 + \cdots + u_{is}x_s.$$

The final step in an ordination by PCA, the step that enables the result of a PCA to be interpreted, is to inspect the pattern of the data points when they are projected onto planes defined by the new, rotated axes (the principal axes). The data points can be projected onto the y_1, y_2 plane, the y_1, y_3 plane, the y_2, y_3 plane, and, indeed, any plane specified by two axes. It is sometimes helpful to look at a perspective drawing or a solid model of pins stuck in cork-board that shows the pattern of the data points relative to three of the principal axes, usually the y_1, y_2, and y_3-axes.

We must now define precisely the consequences of rotating a diffuse, irregular s-dimensional data swarm so as to make its projections onto spaces of fewer dimensions (in practice, onto spaces of two or three dimensions) "as simple as possible" or, perhaps more accurately, "as revealing as possible." Recall that if one does a PCA of the corner points of a box—and it can be an s-dimensional box with s taking any value—the pattern of the points when projected onto any of the two-dimensional planes of the rotated coordinate frame is always a rectangle. What can be said of the projections onto different two-spaces, defined by principal axes, of a diffuse swarm?

The answer is as follows. Consider the new SSCP matrix and the new covariance matrix calculated from the principal component scores after the PCA has been done and the coordinates of the points have been converted from species quantities (xs) into principal component scores (ys). These matrices are

$$\mathbf{R_Y} = \mathbf{YY}' \quad \text{and} \quad \frac{1}{n}\mathbf{R_Y} = \frac{1}{n}\mathbf{YY}',$$

respectively.

It should now be recalled, from Equation (4.3) and Table 4.2 (pages 141 and 142), that both $\mathbf{R_Y}$ and $(1/n)\mathbf{R_Y}$ are diagonal matrices: all their elements except for those on the main diagonal are zero. Now, we already know [see Equation (3.9), page 106] that

$$\frac{1}{n}\mathbf{R_Y} = \begin{pmatrix} \text{var}(y_1) & \text{cov}(y_1, y_2) & \cdots & \text{cov}(y_1, y_s) \\ \text{cov}(y_2, y_1) & \text{var}(y_2) & \cdots & \text{cov}(y_2, y_s) \\ \cdots & \cdots & \cdots & \cdots \\ \text{cov}(y_s, y_1) & \text{cov}(y_s, y_2) & \cdots & \text{var}(y_s) \end{pmatrix}.$$

It therefore follows that the covariances of all pairs of principal component scores are zero; in other words, they are all uncorrelated with one another.

The chief consequence of a PCA, however, is the following. It can be proved (see e.g., Pielou, 1977) that the first principal axis is so oriented as to make the variance of the n first-principal-component scores as great as possible; these are the scores $y_{11}, y_{12}, \ldots, y_{1n}$, the coordinates of the n data points measured along the y_1-axis (the first principal axis). In colloquial terms, this means that the axis is oriented in such a way that when the n data points are projected onto it they have the greatest possible dispersion or "spread."

The second principal axis is so oriented as to make the variance of the n second-principal-component scores (the values $y_{21}, y_{22}, \ldots, y_{2n}$) as great as possible, subject to the restriction that the second axis must be perpendicular to (synonymously, orthogonal to) the first axis. It is found that the first and second-principal-component scores (the y_1s and y_2s) are uncorrelated (have zero covariance).

Likewise, the third principal axis is so oriented as to make the variance of the n third-principal-component scores as great as possible, subject to the restriction that it must be orthogonal to both the first and second axes. This third set of scores is uncorrelated with either of the other two.

And so on, with each succeeding set of scores accounting for as much as possible of the remaining dispersion. All the new variables (the principal component scores) are uncorrelated with one another.

The orientation of the final, sth, principal axis is fixed; it must be orthogonal to the other $s - 1$ principal axes.

Returning to the numerical example in Table 4.3, it is easily found that the covariance matrix of the principal component scores is

$$\frac{1}{n}\mathbf{R}_{\mathbf{Y}} = \frac{1}{n}\mathbf{YY}' = \begin{pmatrix} 211.704 & 0 \\ 0 & 16.295 \end{pmatrix} = \mathbf{\Lambda}.$$

Thus the variances of the principal component scores are equal to the eigenvalues of their covariance matrix.

It is intuitively evident from inspection of Figures 4.4 and 4.5 that the greatest "spread" of the points is in the direction of the y_1-axis, and also that, although the data points show obvious negative correlation when looked at in the frame of the x_1 and x_2-axes (Figure 4.4a), this correlation vanishes when the points are plotted in the frame of the y_1 and y_2-axes (Figure 4.4*b*).

PCA as here described is often used as an ordination method in ecological work. Such an ordination is a "success" if a large proportion of the total dispersion (or scatter) of the data is parallel with the first two or three principal axes; for then this large proportion of the information contained in the original, unvisualizable s-dimensional data swarm can be plotted in two-space or three-space and examined. This is what ordination by PCA sets out to achieve: the data swarm is to be projected onto the two-dimensional or three-dimensional frame (or frames) that most clearly reveals the real pattern of the data. When three axes are retained, as is very often done, the result is shown in print either as a two-dimensional perspective (or isometric) drawing of a three-dimensional graph, or else as a trio of two-dimensional graphs showing the swarm projected onto the y_1, y_2 plane, the y_1, y_3 plane, and the y_2, y_3 plane, respectively.

The statement that such a two- or three-dimensional display of the original s-dimensional data swarm reveals the real pattern of the data is intuitively reasonable, but it is desirable to define more precisely what is meant by "real pattern." The observed abundances of a large number of species co-occurring in an ecological sampling unit are governed by two factors: first, the joint responses of groups of species to persistent features of the environment; second, the "capricious," unrelated responses of a few individual members of a few species to environmental accidents of the sort that occur sporadically, here and there, and have only local and temporary effects. In the present context the joint, related responses of groups of species constitute "real pattern" or "interesting data structure," and the capricious, sporadic responses amount to "noise." (This is not to say that in other contexts, environmental accidents and the noise they produce may not be a researcher's chief interest.) It has been shown by Gauch (1982b) that displaying the results of a PCA, or indeed of any ordination, in only a few dimensions (typically two or three) does more than merely permit an unvisualizable s-dimensional pattern to be visualized; it also suppresses "noise." This is because the first few principal components of the data—those with the largest variances—nearly always reflect the concerted responses of groups of several species. When a group of species (hence numerous individuals) behave in concert, it is unlikely to be the result of localized, temporary "accidents." Moreover, the fact that many species do, indeed, respond in concert to the "important" features of the environment means that the data body as a whole contains redundancies; therefore, the number of coordinate axes needed to display the "interesting structure" of the data is far less than s, the total number of species observed.

To summarize: ordination permits us to profit from the redundancy in field data. Because of redundancy, not much information is lost by representing a swarm of data points in only a few dimensions. And the discarded information (on the disregarded axes along which the variances are small) is mostly noise (Gauch, 1982b).

The method of doing a PCA ordination described in this section can be modified in various ways as shown in the next section. An example of its use in the way previously described may be found in Jeglum, Wehrhahn and Swan (1971). They sampled the vegetation in various communities in the boreal forest of Saskatchewan and ordinated their data and various subsets of it using PCA.

4.3. FOUR DIFFERENT VERSIONS OF PCA

The method given in the preceding section for carrying out a PCA can be modified in one or both of two ways.

First, one can standardize (or rescale) the data by dividing each element in the centered data matrix $\mathbf{X_R}$ by the standard deviation of the elements in its row. The resulting standardizied centered data matrix $\mathbf{Z_R}$ then has as its (i, j)th element

$$\frac{x_{ij} - \bar{x}_i}{\sigma_i},$$

as we saw in Chapter 3 (page 107). The SSCP matrix divided by n [i.e., $(1/n)\mathbf{Z_R Z'_R}$] is the correlation matrix (see Table 3.6, page 112). The PCA is now carried out by doing an eigenanalysis of the correlation matrix instead of the covariance matrix.

The second modification consists in using uncentered data. Instead of analyzing $(1/n)\mathbf{X_R X'_R}$ as was done in Section 4.2, one analyzes $(1/n)\mathbf{XX'}$.

Of course, both these modifications can be made simultaneously. Thus one can analyze the matrix $(1/n)\mathbf{ZZ'}$ in which the (i, j)th element is x_{ij}/σ_i.

Before discussing the advantages and disadvantages of these various versions of PCA, we compare the results they give when applied to a two-dimensional swarm of 10 data points. The coordinates of the points are the columns of Data Matrix #10 given at the top of Table 4.4. In the

TABLE 4.4. FOUR DIFFERENT PCAS OF DATA MATRIX #10.

The data matrix is

$$\mathbf{X} = \begin{pmatrix} 2 & 25 & 33 & 42 & 55 & 60 & 62 & 65 & 92 & 99 \\ 20 & 30 & 13 & 30 & 17 & 42 & 27 & 25 & 25 & 43 \end{pmatrix}.$$

Unstandardized Uncentered PCA

$$\frac{1}{n}\mathbf{XX}' = \begin{pmatrix} 3644 & 1641 \\ 1641 & 889 \end{pmatrix}$$

$$\mathbf{U} = \begin{pmatrix} 0.906 & 0.423 \\ -0.423 & 0.906 \end{pmatrix}$$

$$\Lambda = \begin{pmatrix} 4409 & 0 \\ 0 & 124 \end{pmatrix}$$

(Axes y_1', y_2' in Figure 4.6*a*)

Unstandardized Centered PCA[a]

$$\frac{1}{n}\mathbf{X}_\mathbf{R}\mathbf{X}_\mathbf{R}' = \begin{pmatrix} 781.9 & 132.3 \\ 132.3 & 93.8 \end{pmatrix}$$

$$\mathbf{U} = \begin{pmatrix} 0.983 & 0.183 \\ -0.183 & 0.983 \end{pmatrix}$$

$$\Lambda = \begin{pmatrix} 806.4 & 0 \\ 0 & 69.2 \end{pmatrix}$$

(Axes y_1, y_2 in Figure 4.6*a*)

Standardized Uncentered PCA

$$\frac{1}{n}\mathbf{ZZ}' = \begin{pmatrix} 4.66 & 6.06 \\ 6.06 & 9.48 \end{pmatrix}$$

$$\mathbf{U} = \begin{pmatrix} 0.561 & 0.828 \\ -0.828 & 0.561 \end{pmatrix}$$

$$\Lambda = \begin{pmatrix} 13.593 & 0 \\ 0 & 0.548 \end{pmatrix}$$

(Axes y_1'', y_2'' in Figure 4.6*b*)

Standardized Centered PCA

$$\frac{1}{n}\mathbf{Z}_\mathbf{R}\mathbf{Z}_\mathbf{R}' = \begin{pmatrix} 1 & 0.488 \\ 0.488 & 1 \end{pmatrix}$$

$$\mathbf{U} = \begin{pmatrix} 0.707 & 0.707 \\ -0.707 & 0.707 \end{pmatrix}$$

$$\Lambda = \begin{pmatrix} 1.488 & 0 \\ 0 & 0.512 \end{pmatrix}$$

(Axes y_1''', y_2''' in Figure 4.6*b*)

[a]This is the version of PCA described and demonstrated in Section 4.2.

separate sections in the lower part of the table are given, for each of the four forms of PCA: (1) the square symmetric matrix to be analyzed, (2) the matrix of eigenvectors **U**, and (3) the matrix of eigenvalues **Λ**.

The results are shown graphically in Figure 4.6. It should be noticed that the effect of standardizing the raw data (as in Figure 4.6*b*) is to make the variances of both sets of coordinates equal to unity. Thus the dispersions of the points along the x_1/σ_1 axis and along the x_2/σ_2 axis are the same. Standardizing the data, therefore, alters the shape of the swarm; after standardization, the swarm is noticeably less elongated than it was before.

PCA Using a Correlation Matrix

Analysis of the correlation matrix $(1/n)\mathbf{Z}_\mathbf{R}\mathbf{Z}_\mathbf{R}'$ is a form of PCA that is frequently encountered in the ecological literature. As has already been explained, the correlation matrix is obtained from the standardized centered

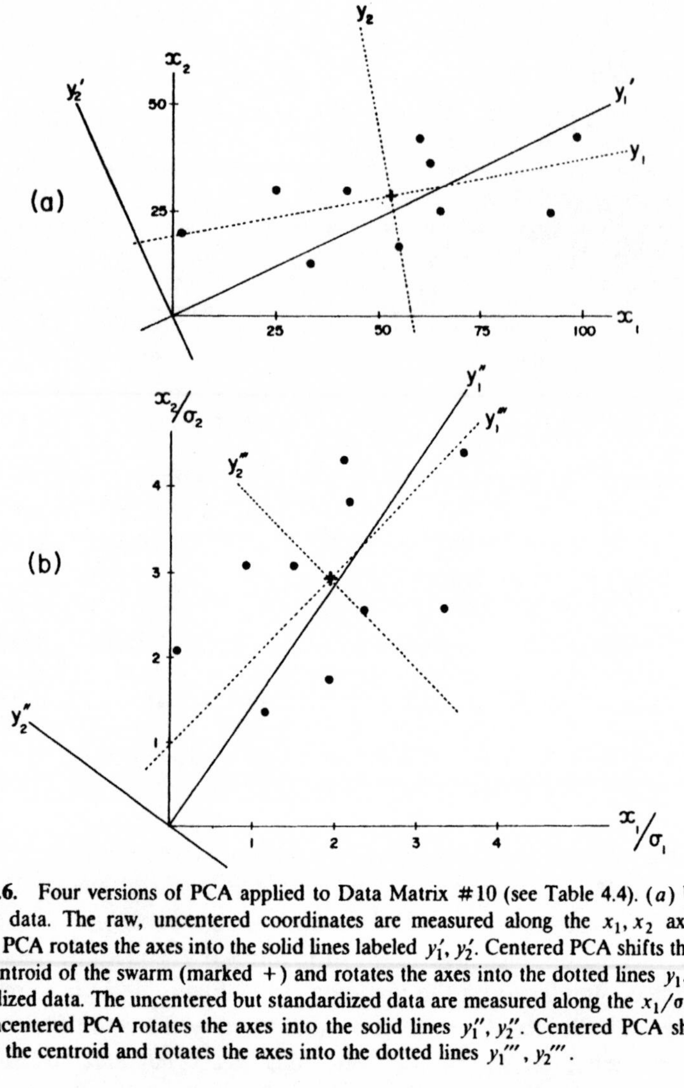

Figure 4.6. Four versions of PCA applied to Data Matrix #10 (see Table 4.4). (*a*) Unstandardized data. The raw, uncentered coordinates are measured along the x_1, x_2 axes. Uncentered PCA rotates the axes into the solid lines labeled y_1', y_2'. Centered PCA shifts the origin to the centroid of the swarm (marked +) and rotates the axes into the dotted lines y_1, y_2. (*b*) Standardized data. The uncentered but standardized data are measured along the $x_1/\sigma_1, x_2/\sigma_2$ axes. Uncentered PCA rotates the axes into the solid lines y_1'', y_2''. Centered PCA shifts the origin to the centroid and rotates the axes into the dotted lines y_1''', y_2'''.

data matrix $\mathbf{Z_R}$. In what follows we discuss the merits of standardization taking it for granted that the data have first been centered by rows. The pros and cons of data centering is discussed in a subsequent subsection.

In some analyses standardization of the data is a (possibly) desirable option; in others it is a necessity. We consider these contrasted situations in turn.

Standardization is often desirable as a way of preventing the "swamping" of the uncommon species in a community by the common or abundant ones. Unless data are standardized, the dominant species are likely to dominate the analysis. This happens because the quantities of abundant species tend to have higher variances (as well as higher means) than the quantities of uncommon species. Standardization equalizes all the variances before axis rotation (the analysis itself) is carried out. Thus if one wishes subordinate species to have an appreciable effect on the outcome, it is a good idea to use standardized data.

However, this does not mean that standardization is always desirable. It is a matter of judgment. It could well be argued that the dominant species ought to control the result simply because they are dominant. Further, there is a risk that standardization may give rare species an undesirable prominence; if their presence is due only to chance, and is not a response to an environmental variable of interest, they are merely "noise." Therefore, deciding whether to standardize or not to standardize entails a trade-off between underemphasizing and overemphasizing the less common species. A useful compromise is to exclude truly rare species from the raw data, and after that to standardize these edited data for analysis (Nichols, 1977).

Standardization of the data must be done when the quantities of the different species are measured in different units. In vegetation studies, for instance, it is often convenient to record the amounts of some species by counting individuals and of other species by measuring cover. When this is done, the species quantities are obviously noncomparable in their raw form and should be standardized before an analysis is done.

Data matrices whose elements are the values of noncomparable environmental variables should also be standardized. Let us examine a particular example in some detail.

Newnham (1968) did a PCA of a correlation matrix that showed the pairwise correlations among 19 climatic variables measured at 70 weather stations in British Columbia. Thus the data matrix had 19 rows in which were recorded such variables as average daily minimum temperature in winter, average winter precipitation, average length of frost-free period and the like, and 70 columns, one for each weather station. An eigenanalysis of the 19×19 correlation matrix, whose elements were the correlations between every possible pair (171 pairs) of climatic variables yielded 19 eigenvectors. If we write $(u_{i,1}, u_{i,2}, \ldots, u_{i,19})$ for the ith eigenvector, then the ith principal component is

$$y_i = u_{i,1}z_1 + u_{i,2}z_2 + \cdots + u_{i,19}z_{19}$$

where the zs are the elements of the original data matrix after they have been centered and standardized. Newnham scaled the eigenvector elements (the us) so that the largest element of each was equal to unity (it is the relative, not the absolute, magnitudes of the elements of an eigenvector that matter, and one can choose whatever scale happens to be convenient).

The first two principal components—those with the largest eigenvalues—were as follows. Only the five terms with the largest coefficients ("weights") are shown here:

$$y_1 = 0.97z_3 + z_7 + 0.99z_{10} + z_{11} + 0.99z_{14} + 14 \text{ other terms;}$$

$$y_2 = z_4 + 0.99z_5 + 0.94z_{13} - 0.73z_{16} - 0.86z_{17} + 14 \text{ other terms.}$$

The five most heavily weighted variables contributing to these two principal components are as follows:

Contributing to y_1:

z_3 winter temperature: average daily maximum, °F;
z_7 winter temperature: average daily minimum, °F;
z_{10} fall temperature: average daily minimum, °F;
z_{11} winter temperature: average daily mean, °F;
z_{14} fall temperature: average daily mean, °F.

Contributing to y_2:

z_4 spring temperature: average daily maximum, °F;
z_5 summer temperature: average daily maximum, °F;
z_{13} summer temperature: average daily mean, °F;
z_{16} spring precipitation: average in inches;
z_{17} summer precipitation: average in inches.

These two components accounted for 57.4% and 29.4% of the variance in the data, respectively, for a total of 86.8%.

When we consider the variables that are weighted most heavily in the first two principal components, it is seen that the weather stations (the data points) where winters and falls are mild have the highest first principal component scores, and the stations where springs and summers are hot and dry (notice the negative coefficients of z_{16} and z_{17}) have the highest second principal component scores. Thus we can draw the two-dimensional coordinate frame shown in Figure 4.7*a* and label the four regions into which the axes divide it with a two-sentence description of the climate: the first sentence puts into words the meaning of high and low values of the first principal component, and analogously for the second sentence.

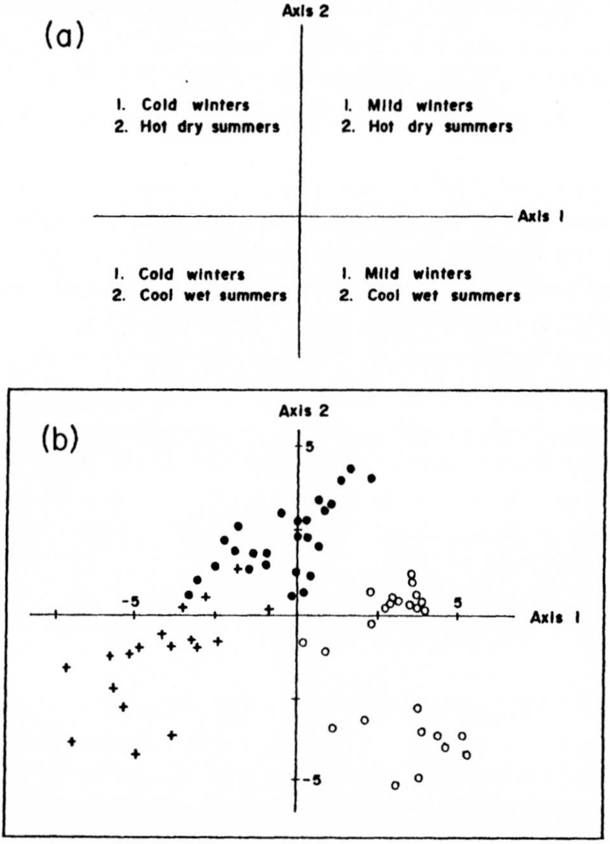

Figure 4.7. (*a*) A "qualitative" graph labeling the regions in the ordination of 70 British Columbia weather stations shown in (*b*). The first two axes obtained from a PCA of climatic data divide the coordinate frame into four regions. Axis 1 separates the stations into those with mild and those with cold winters. Axis 2 separates the stations into those with hot dry summers and those with cool wet summers. (*b*) A two-dimensional ordination, by PCA of the correlation matrix, of the 70 weather stations. The symbol ● denotes a station where ponderosa pine occurs; O denotes sitka spruce; +, neither species. The two species are never found together. (Adapted from Newnham, 1968.)

The scatter diagram in Figure 4.7*b* is a two-dimensional ordination of the 70 weather stations. Each station is represented by a point having its first and second principal component scores as coordinates. Three different symbols have been used for the points: one for stations where Sitka spruce (*Picea sitchensis*) occurs; one for stations where ponderosa pine (*Pinus ponderosa*) occurs; one for stations where neither tree species is found. As one would expect from a knowledge of these trees' habitat requirements and geographic ranges, Sitka spruce occurs predominantly at stations with marine climates (mild winters and cool, wet summers) and ponderosa pine at interior stations with hot, dry summers.

The figure demonstrates very well how an ordination by PCA can clarify what was originally a confusing and unwieldy 19×70 data matrix. Two coordinate axes (instead of 19) suffice to portray a large proportion (86.8%) of the information in the original data, and concrete meanings can be attached to the scores measured along each axis.

This last point deserves strong emphasis. Ordinations, especially of community data, are often presented in the ecological literature with the axes cryptically labeled "Axis 1," "Axis 2," and so on, with no explanation as to the concrete meaning, the actual ecological implications, of these coordinate axes. Without such explanations the scatter diagrams yielded by ordinations are uninterpretable. As Nichols (1977) has written, "The primary effort in any PCA should be the examination of the eigenvector coefficients [to] determine which species [or environmental variables] combine to define which axes, and why."

PCA Using Uncentered Data

The great majority of ecological ordinations are done with centered data but this is not always the most appropriate procedure. Sometimes it is preferable to ordinate data in their raw, uncentered form. The reason for this will not become clear until we consider an example with more than two species, and hence a data swarm occupying more than two dimensions. First, however, it is worth looking at the results of doing both a centered and an uncentered PCA on the same, deliberately simplified, two-dimensional data swarm chosen to demonstrate as clearly as possible the contrast between the two methods.

Consider Figure 4.8. Both graphs show the same seven data points plotted in raw form in the frame defined by the x_1 and x_2-axes. The original

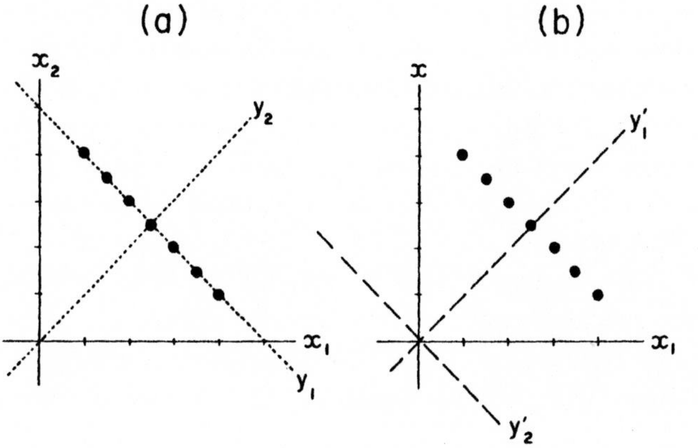

Figure 4.8. Both graph show a row of seven data points plotted in an x_1, x_2 coordinate frame. (a) The dotted lines y_1 and y_2 are the first and second principal axes of a centered PCA; (b) the dashed lines y_1' and y_2' are the first and second principal axes of an uncentered PCA.

data matrix is

$$\mathbf{X} = \begin{pmatrix} 8 & 7 & 6 & 5 & 4 & 3 & 2 \\ 2 & 3 & 4 & 5 & 6 & 7 & 8 \end{pmatrix}.$$

Figure 4.8a shows the principal axes (the lines y_1 and y_2) yielded by a centered PCA. The intersection of these axes, which is the origin of the new frame, is at the centroid of the swarm (which coincides with the central point of the seven). The coordinates of the data points relative to the y_1 and y_2-axes are given by the matrix

$$\mathbf{Y} = \begin{pmatrix} -4.24 & -2.83 & -1.41 & 0 & 1.41 & 2.83 & 4.24 \\ 0 & 0 & 0 & 0 & 0 & 0 & 0 \end{pmatrix}.$$

Clearly, the y_1-axis is so aligned that the points have the maximum possible "spread" along it; their spread relative to the y_2-axis is zero.

Figure 4.8b shows the principal axes (the lines y_1' and y_2') yielded by an uncentered PCA. The intersection of the new axes coincides with the intersection of the old axes; equivalently, the origin has not been shifted. The coordinates of the data points relative to the y_1' and y_2'-axes are given by

$$\mathbf{Y}^{(\prime)} = \begin{pmatrix} 7.07 & 7.07 & 7.07 & 7.07 & 7.07 & 7.07 & 7.07 \\ -4.24 & -2.83 & -1.41 & 0 & 1.41 & 2.83 & 4.24 \end{pmatrix}.$$

(The prime is in parentheses to show that it does not indicate a transposed matrix.) The y_1'-axis is so aligned that the sum of squares of the y_1' coordinates (their first-principal-component scores) is as great as possible. It is $7 \times 7.071^2 = 350$. Any other direction for y_1' would give a smaller value. The spread of the points *along* the y_1'-axis is irrelevant (it is zero in the example). It is in this that the contrast between a centered and an uncentered PCA consists.

It should also be noticed that the data points' second-principal-component scores, given by their projections onto the y_2'-axis, are the same (in this geometrically regular case) as the first-principal-component scores yielded by the centered PCA. It often happens, with real, many-dimensional data, that that the second, third, fourth,... principal axes of an uncentered PCA are roughly parallel with (hence give roughly the same scores as) the first, second, third,... principal axes of a centered PCA; the equality is never exact except in deliberately contrived, geometrically regular data such as that in Figure 4.8. And it does not always happen, as we shall see.

It is worth reemphasizing the contrast between the two analyses. In both cases the first axis was aligned so as to maximize the sum of squares of the seven points' coordinates measured along these axes. These sums of squares are

$$\sum_{i=1}^{7} (y_{1i})^2 \quad \text{and} \quad \sum_{i=1}^{7} (y_{1i}')^2$$

in the centered and uncentered cases, respectively. With a centered PCA, *because* it is centered, this sum of squares is proportional to the variance of the seven y_{1i} values. With an uncentered PCA this sum of squares is the sum of squared distances of the points (after projection onto the y_1'-axis) from the unshifted origin; it bears no relation to the variance of the y_{1i}' values, which in the example is zero.

We now turn to a three-dimensional example to illustrate the ecological usefulness of an uncentered PCA. In practice, of course, there are usually far more than three axes (three species) and it is unfortunate that we must limit ourselves to three dimensions in order to make the analysis visualizable. The reader should find it easy to extrapolate the arguments to the many-dimensional case.

The three-dimensional example is shown in Figure 4.9. The points in the three-dimensional scatter diagram at the top of the figure clearly belong to

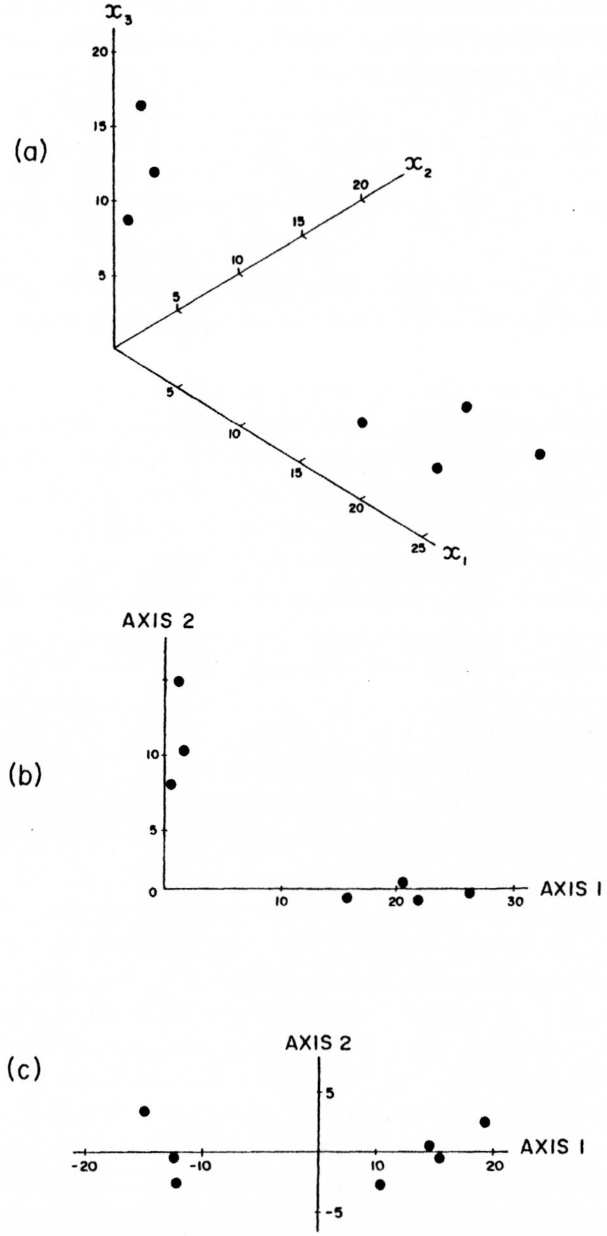

Figure 4.9. (*a*) Seven data points in three-space. They form two qualitatively different clusters. (*b*) A plot of the points in the coordinate frame formed by the first two principal axes resulting from an uncentered PCA. One cluster lies on axis 1 and the other very close to axis 2. (*c*) The corresponding plot after a centered PCA. Both clusters lie on axis 1.

two qualitatively different clusters; there is evidently a four-member set of quadrats (cluster 1) containing only species 1 and 2, and a three-member set (cluster 2) containing species 3 together with small amounts of species 2. The data matrix (Data Matrix #11) is

$$\mathbf{X} = \begin{pmatrix} 15 & 18 & 21 & 24 & 0 & 0 & 0 \\ 5 & 10 & 5 & 10 & 1 & 2 & 3 \\ 0 & 0 & 0 & 0 & 8 & 15 & 10 \end{pmatrix}.$$

As may be seen, the axes yielded by an uncentered PCA (Figure 4.9*b*) could be said to "define" the two contrasted clusters. The first axis transfixes one of the clusters and the second axis grazes the other.

With a centered PCA, the first axis goes through both clusters (Figure 4.9*c*). There is a perceptible gap between the clusters but their qualitative dissimilarity (cluster 1 lacks species 3, and cluster 2 lacks species 1) is not nearly so well brought out.

It should now be clear that in certain circumstances an uncentered PCA reveals the structure of the data more clearly than does a centered PCA.

An uncentered PCA is called for when the data exhibit *between-axes heterogeneity*, that is, when there are clusters of data points such that each cluster has zero (or negligibly small) projections on some subset of the axes, a different subset of axes for each cluster. Thus in the example in Figure 4.9, cluster 1 has zero projection on the x_3-axis, and cluster 2 has zero projection on the x_1-axis. When an uncentered PCA is done on data of this kind, each of the first few principal axes passes through (or very close to) one of the qualitatively different clusters. Moreover, these axes tend to be *unipolar*. On a unipolar axis all the data points have scores of the same sign, all positive or all negative. In the example, as can be seen in Figure 4.9*b*, all the points of cluster 1 have positive projections on axis 1; likewise, all the points of cluster 2 have positive projections on axis 2.

A centered PCA is called for when the data exhibit little or no between-axes heterogeneity and nearly all the heterogeneity in the data is *within-axes heterogeneity* or, equivalently, when the data points have appreciable projections on all axes. With a centered PCA all the principal axes are *bipolar*: on each of them some of the data points have positive scores and some negative scores. Axes 1 and 2 in Figure 4.9*c* are both bipolar.

Putting these requirements into ecological terms, it is seen that an uncentered ordination is called for when the quadrats belong to groups having nonidentical lists of common species. A centered PCA is called for when the contrast among the quadrats is less pronounced and their contents differ in degree rather than in kind.

In practice data are often obtained for which it is not immediately obvious whether the between-axes heterogeneity exceeds the within-axes heterogeneity or vice versa. When this happens, it is best to do both a centered and an uncentered PCA. If the between-axes heterogenity of the data is appreciable, then there will be as many unipolar (or almost unipolar, see later) axes as there are qualitatively different clusters of data points. Of course, the first axis of an uncentered PCA is automatically unipolar, regardless of whether there is any between-axes heterogenity. If there is not, then the first axis is merely a line through the origin of the raw coordinate frame passing close to the centroid of the whole data swarm as in Figure 4.6 (page 154), for instance.

Data are often obtained that do not clearly belong to one type or the other. Then an uncentered PCA is likely to give one or more principal axes (after the first) that, although technically bipolar, are so "unsymmetrical" that it seems reasonable to treat them as "virtually" unipolar. A bipolar axis is said to be symmetrical if, relative to it, the totals of the positive and negative scores are equal. Obviously, bipolar axes can range from the perfectly symmetrical to the strongly asymmetrical; only in the limit, when the asymmetry becomes total (all scores of the same sign), is an axis unipolar. Therefore, in ecological contexts an axis need not be strictly unipolar to suggest the existence of qualitatively different clusters within a body of data. Noy-Meir (1973a) has devised a *coefficient of asymmetry* for principal axes that ranges from 0 (for perfect symmetry) to 1 (for complete asymmetry). He recommends that any axis for which the coefficient of asymmetry exceeds 0.9 be regarded as virtually unipolar.

Given an uncentered PCA, the coefficient, α, is defined as follows.* Let the elements of the eigenvector defining an axis be denoted by us; let u_+ denote a positive element and u_- a negative element. Then the coefficient is

$$\alpha = 1 - \frac{\sum u_-^2}{\sum u_+^2} \qquad \left(\text{if } \sum u_-^2 < \sum u_+^2\right)$$

*This definition is not applicable to the axes of a centered PCA since the data points themselves have negative as well as positive coordinates. As α values are only of interest for uncentered axes, there is no point in adapting the formula for α to make it applicable to centered axes.

or

$$\alpha = 1 - \frac{\sum u_+^2}{\sum u_-^2} \qquad \left(\text{if } \sum u_+^2 < \sum u_-^2\right).$$

Let us find the coefficients of asymmetry for axes 1 and 2 in Figure 4.9*b*. An uncentered (and unstandardized) PCA of Data Matrix #11 yields a matrix of eigenvectors

$$\mathbf{U} = \begin{pmatrix} 0.931 & 0.364 & 0.018 \\ -0.078 & 0.151 & 0.985 \\ -0.356 & 0.919 & -0.169 \end{pmatrix}.$$

Therefore, $\alpha = 1$ for axis 1, since all elements in the first row of **U** are of the same sign. This result is also obvious from the figure, of course.

For axis 2,

$$\alpha = 1 - \frac{(-0.078)^2}{0.151^2 + 0.985^2} = 0.994.$$

It is clear from Figure 4.9*b* that axis 2 should be treated as unipolar even though three of the points belonging to the cluster on axis 1 have small negative scores with respect to axis 2. It is these small negative scores that cause α to be just less than 1. Using Noy-Meir's criterion whereby a value of α greater than 0.9 is treated as indicating a virtually unipolar axis, we may treat axis 2 in Figure 4.9*b* as unipolar.

A clear and detailed discussion of the use of centered and uncentered ordinations on different kinds of data has been given by Noy-Meir (1973a).

For an example of the practical application of uncentered PCA to field data, see Carleton and Maycock (1980). These authors used the method to identify "vegetational noda" (qualitatively different communities) in the boreal forests of Ontario south of James Bay.

Other Forms of PCA

There are other ways (besides those described in the preceding pages) in which data can be transformed as a preliminary to carrying out a PCA. Data can be standardized in various different ways, and they can be centered in various different ways. Standardizing and centering can be done separately or in combination. There are numerous possibilities.

Most of the methods are seldom used by ecologists and are not dealt with in this book. They are clearly discussed and compared in Noy-Meir (1973a) and Noy-Meir, Walker, and Williams (1975).

4.4. PRINCIPAL COORDINATE ANALYSIS

The methods of ordination so far discussed (the various versions of PCA) all operate on a swarm of data points in s-space and specify different ways of projecting these points onto a space of fewer than s dimensions. The origin may be shifted and the scales of the axes may be changed, but at the outset each data point has, as its coordinates, the amount of each species in the quadrat represented by the point. Principal *coordinate* analysis differs from principal *component* analysis in the way in which the data swarm is constructed to begin with. The points are not plotted in an s-dimensional coordinate frame. Instead, their locations are fixed as follows: the dissimilarity between every pair of quadrats is measured, using some chosen measure of dissimilarity, and the points are then plotted in such a way as to make the distance between every pair of points as nearly as possible equal to their dissimilarity. It should be noticed that the number of axes of the coordinate frame in which the points are plotted depends on the number of points, not on the number of species. Also, that the value of the coordinates are of no intrinsic interest; they merely ensure that the points shall have the desired spacing.

Before describing how the coordinates are found, we need an acronym for "principal coordinate analysis." None has come into common use. In what follows it is called PCO.

To do a PCO, a measure of interquadrat dissimilarity must first be chosen. Any metric measure may be used. Without specifying the measure, let us write $\delta(j, k)$ for the dissimilarity between quadrats j and k. We require to find coordinates for n points in n-space such that the distance between points j and k, namely, $d(j, k)$, shall be equal (or as nearly equal as possible) to $\delta(j, k)$. Often it proves impossible to arrange the points so that their pairwise distances have exactly the required values and one must be content with approximate equalities.

It should be noticed that a space of $n - 1$ dimensions always suffices to contain n points. Thus two points can always be contained in a one-space (a line); three points can always be contained in a two-space (a plane), or in a one-space if they chance to be colinear; similarly, n points can always be contained in a space of $n - 1$ dimensions at most. Hence it could be said that we need to find the coordinates of the n points in $(n - 1)$-space rather than in n-space. This is, indeed, true but the argument is simpler and clearer if an n-space is considered. The required $(n - 1)$-space is a subspace of this n-space (in the same way that a two-dimensional floor is a subset of a three-dimensional room). Equivalently, all the points have a coordinate of zero on one of the n axes, the same axis for all of them. There is no need to keep this fact in mind; the required zeros emerge as part of the output of the computations.

It could be argued that PCO is necessarily inferior to PCA because in PCA each point is placed exactly where it "ought" to be, whereas in PCO each point is so placed that interpoint distances are as closely as possible (but seldom exactly) equal to interquadrat distances. Provided the approximation is close, the imprecision of PCO is of no practical consequence. With both PCO and PCA, the final step consists in projecting the swarm to be examined onto a visualizable space of two or three dimensions, and far more information is usually lost in this reduction of dimensionality than in slight "misplacings" of the points in the n-dimensional swarm. However, PCO is not suitable for all bodies of data. We consider later (after describing the method) how to judge when PCO is appropriate.

Now for the method. Any metric measure of dissimilarity may be used. The object is to find coordinates for n points in n-space such that $d(j, k)$, the distance between points j and k, shall be as nearly equal as possible to $\delta(j, k)$, their dissimilarity, however we have chosen to measure it.

For clarity, the argument (which is rather long) is given in the following numbered paragraphs. After that, the operations to be performed are briefly reiterated in recipe form. Readers who wish to try the method before delving into the reasoning that underlies it should skip to the recipe and return to the details later.

1. As a preliminary, note that all summations are over the range 1 to n. Therefore, to keep the symbols as uncluttered as possible, these limits are left unstated. However, it is very important to observe *which* of the subscripts varies each time a summation is done. It is specified by the symbol below the Σ. Bear in mind, for example, that a sum such as $\Sigma_r c_{rj} = c_{1j} + c_{2j} + \cdots + c_{nj}$ (in which r takes the series of values

$1, 2, \ldots, n$) is not the same as $\Sigma_j c_{rj} = c_{r1} + c_{r2} + \cdots + c_{rn}$ (in which j takes the series of values $1, 2, \ldots, n$).

2. The coordinates sought are to be written as an $n \times n$ matrix **C** in which each column gives the n coordinates of one of the points. The points are to be centered. That is, the origin of the coordinates is to be at the centroid of the swarm of points. Equivalently, $\Sigma_j c_{rj} = 0$ for all r.

3. The distance², $d^2(j, k)$, between the jth and kth points is

$$d^2(j,k) = \sum_r (c_{rj} - c_{rk})^2.$$

Here r denotes the rth row of matrix **C**. Each row of **C** corresponds with an axis in the n-space that is to contain the swarm of points but (in contrast to PCA) the axes do not represent species.

4. It follows that

$$d^2(j,k) = \sum_r \left(c_{rj}^2 + c_{rk}^2 - 2c_{rj}c_{rk}\right)$$

$$= \sum_r c_{rj}^2 + \sum_r c_{rk}^2 - 2\sum_r c_{rj}c_{rk}.$$

5. Next consider the $n \times n$ matrix, say **A**, formed by premultiplying **C** by its transpose **C′**. It is

$$\mathbf{A} = \mathbf{C'C} = \begin{pmatrix} c_{11} & c_{21} & \cdots & c_{n1} \\ c_{12} & c_{22} & \cdots & c_{n2} \\ \cdots & \cdots & \cdots & \cdots \\ c_{1n} & c_{2n} & \cdots & c_{nn} \end{pmatrix} \begin{pmatrix} c_{11} & c_{12} & \cdots & c_{1n} \\ c_{21} & c_{22} & \cdots & c_{2n} \\ \cdots & \cdots & \cdots & \cdots \\ c_{n1} & c_{n2} & \cdots & c_{nn} \end{pmatrix}$$

$$= \begin{pmatrix} \sum_r c_{r1}^2 & \sum_r c_{r1}c_{r2} & \cdots & \sum_r c_{r1}c_{rn} \\ \sum_r c_{r2}c_{r1} & \sum_r c_{r2}^2 & \cdots & \sum_r c_{r2}c_{rn} \\ \cdots & \cdots & \cdots & \cdots \\ \sum_r c_{rn}c_{r1} & \sum_r c_{rn}c_{r2} & \cdots & \sum_r c_{rn}^2 \end{pmatrix}.$$

A is obviously symmetrical. In the following paragraphs we first find the elements of **A** from the interquadrat dissimilarities which are calculated from the field observations. We then find the elements of **C** from those of **A**.

6. It follows from paragraph 5 that if we write a_{jk} for the (j, k)th element of **A**, we have

$$a_{jk} = \sum_r c_{rj} c_{rk}; \qquad a_{jj} = \sum_r c_{rj}^2; \qquad a_{kk} = \sum_r c_{rk}^2.$$

Therefore, an alternative formula for $d^2(j, k)$ is

$$d^2(j,k) = a_{jj} + a_{kk} - 2a_{jk}. \tag{4.4}$$

7. Notice, for later use, that $\Sigma_j a_{jk} = 0$. This follows from the fact that

$$\sum_j a_{jk} = \sum_j \left(\sum_r c_{rj} c_{rk} \right) = \sum_r c_{rk} \left(\sum_j c_{rj} \right) = 0$$

since the sum $\Sigma_j c_{rj}$ is zero (see paragraph 2). Because **A** is symmetric, it is also true that $\Sigma_k a_{jk} = 0$.

8. We now wish to find a_{jk} as a function of $d^2(j, k)$. Rearranging Equation (4.4) shows that

$$a_{jk} = \tfrac{1}{2}\left[-d^2(j,k) + a_{jj} + a_{kk} \right]. \tag{4.5}$$

9. We now find a_{jj} and a_{kk} as functions of $d^2(j, k)$. To do this, sum every term in (4.4) over all values of j, from 1 to n. Thus

$$\sum_j d^2(j,k) = \sum_j a_{jj} + \sum_j a_{kk} - 2\sum_j a_{jk}.$$

Put $\Sigma_j a_{jj} = x$; note that $\Sigma_j a_{kk} = n a_{kk}$, and recall from paragraph 7 that $\Sigma_j a_{jk} = 0$. Therefore,

$$\sum_j d^2(j,k) = x + n a_{kk} \quad \text{whence} \quad a_{kk} = \frac{1}{n}\left(\sum_j d^2(j,k) - x \right).$$

Likewise, summing every term in (4.4) over all values of k, it is seen that

$$\sum_k d^2(j,k) = n a_{jj} + x \quad \text{whence} \quad a_{jj} = \frac{1}{n}\left(\sum_k d^2(j,k) - x \right).$$

Observe that $\Sigma_k a_{kk} = \Sigma_j a_{jj}$, which we have already denoted by x.

Equation (4.5) thus becomes

$$a_{jk} = \frac{1}{2}\left\{-d^2(j,k) + \frac{1}{n}\left(\sum_j d^2(j,k) - x\right) + \frac{1}{n}\left(\sum_k d^2(j,k) - x\right)\right\}$$

$$= -\frac{1}{2}d^2(j,k) + \frac{1}{2n}\sum_j d^2(j,k) + \frac{1}{2n}\sum_k d^2(j,k) - \frac{x}{n} \qquad (4.6)$$

and it remains to express x as a function of $d^2(j,k)$.

10. In paragraph 6 it was shown that

$$a_{jj} = \sum_r c_{rj}^2.$$

Therefore,

$$\sum_j a_{jj} = \sum_j \sum_r c_{rj}^2. \qquad (4.7)$$

Recall from paragraph 1 that the centroid of the n points whose coordinates we seek is to be at the origin. Therefore, $\Sigma_r c_{rj}^2$ is the square of the distance of the jth point from the origin, and $\Sigma_j \Sigma_r c_{rj}^2$ is the sum of the squares of the distances of all n points from the origin. This latter sum is equal* to $(1/n)$ times the sum of the squares of the distances between every pair of points. That is,

$$\sum_j \sum_r c_{rj}^2 = \frac{1}{n}\sum_{j<k} d^2(j,k) = \frac{1}{2n}\sum_j \sum_k d^2(j,k).$$

(The form $\Sigma_{j<k}$ specifies that each pair of points shall be considered only once. The form $\Sigma_j \Sigma_k$ specifies that $d^2(j,k)$ and $d^2(k,j)$ which is the same, shall both enter the sum; hence the 2 in the denominator.)

11. Equation (4.6) now becomes

$$a_{jk} = -\frac{1}{2}d^2(j,k) + \frac{1}{2n}\sum_j d^2(j,k)$$

$$+ \frac{1}{2n}\sum_k d^2(j,k) - \frac{1}{2n^2}\sum_j \sum_k d^2(j,k).$$

*This fact was used in another context earlier. A proof is found in Pielou (1977), p. 320.

12. We now stipulate that $d^2(j,k)$ (the distance[2] between points j and k) is to be equal (or as nearly equal as possible) to $\delta^2(j,k)$. We therefore put

$$a_{jk} = -\frac{1}{2}\delta^2(j,k) + \frac{1}{2n}\sum_j \delta^2(j,k) + \frac{1}{2n}\sum_k \delta^2(j,k) - \frac{1}{2n^2}\sum_j\sum_k \delta^2(j,k) \quad (4.8)$$

The numerical values of $\delta^2(j,k)$ have been calculated for every pair of points j and k. Hence all the elements of **A** can be given numerical values. It remains to find the elements of **C**.

13. Since **A** is a square symmetric matrix, we know that

$$\mathbf{A} = \mathbf{U}'\boldsymbol{\Lambda}\mathbf{U},$$

where $\boldsymbol{\Lambda}$ is the diagonal matrix whose nonzero elements are the eigenvalues of **A**, and **U** is the orthogonal matrix whose rows are the corresponding eigenvectors of **A**; **U′** is the transpose of **U**.

Since $\boldsymbol{\Lambda}$ is a diagonal matrix, it can be replaced by the product $\boldsymbol{\Lambda}^{1/2}\boldsymbol{\Lambda}^{1/2}$ in which the nonzero elements are $\lambda_1^{1/2}, \lambda_2^{1/2}, \ldots, \lambda_n^{1/2}$.

Thus

$$\mathbf{A} = (\mathbf{U}'\boldsymbol{\Lambda}^{1/2})(\boldsymbol{\Lambda}^{1/2}\mathbf{U}).$$

Each of these factors is the transpose of the other.

Now recall (from paragraph 1) that, by definition,

$$\mathbf{A} = \mathbf{C}'\mathbf{C}.$$

Therefore,

$$\mathbf{U}'\boldsymbol{\Lambda}^{1/2} = \mathbf{C}' \quad \text{and} \quad \boldsymbol{\Lambda}^{1/2}\mathbf{U} = \mathbf{C}.$$

14. To find the elements of **C**, we therefore carry out an eigenanalysis of **A**. Then the first principal coordinates of the data points, which are the elements in the first row of **C**, are obtained from $\lambda_1^{1/2}\mathbf{u}_1'$ ($\mathbf{u}_1'$ is the first eigenvector of **A**). The second principal coordinates are obtained from $\lambda_2^{1/2}\mathbf{u}_2'$, and so on. If, as is very often the case, a two-dimensional ordination

is wanted, only the first two rows of **C** need be evaluated; they give the coordinates of the n points in two-space with the points so arranged that the distance between every pair of points approximates as closely as possible their dissimilarity as calculated at the outset.

To reiterate, without explanations: the operations required to do a PCO are the following.

1. Calculate the dissimilarity between every pair of quadrats, using some chosen dissimilarity measure. Denote by $\delta(j,k)$ the dissimilarity between quadrats j and k. Put the squares of these dissimilarities as the elements of an $n \times n$ matrix $\boldsymbol{\Delta}$.
2. Find the elements of the $n \times n$ symmetric matrix **A** from Equation (4.8) which is

$$a_{jk} = -\frac{1}{2}\delta^2(j,k) + \frac{1}{2n}\sum_j \delta^2(j,k) + \frac{1}{2n}\sum_k \delta^2(j,k) - \frac{1}{2n^2}\sum_j\sum_k \delta^2(j,k)$$

$\Sigma_j \delta^2(j,k)$ is the sum of the elements in the jth row of $\boldsymbol{\Delta}$;
$\Sigma_k \delta^2(j,k)$ is the sum of the elements in the kth column of $\boldsymbol{\Delta}$;
$\Sigma_j \Sigma_k \delta^2(j,k)$ is the sum of all the elements in $\boldsymbol{\Delta}$.

A more compact formula for determining **A** from $\boldsymbol{\Delta}$ is given in Exercise 4.6.

3. Do an eigenanalysis of **A**.
4. For a two-dimensional PCO, calculate the first two rows of **C** which are

$$(c_{11} \quad c_{12} \quad \cdots \quad c_{1n}) = \lambda_1^{1/2}(u_{11} \quad u_{12} \quad \cdots \quad u_{1n})$$

and

$$(c_{21} \quad c_{22} \quad \cdots \quad c_{2n}) = \lambda_2^{1/2}(u_{21} \quad u_{22} \quad \cdots \quad u_{2n}).$$

Here λ_1 and λ_2 are the first and second eigenvalues of **A**; $(u_{11} \; u_{12} \; \cdots \; u_{1n})$ and $(u_{21} \; u_{22} \; \cdots \; u_{2n})$ are the respective eigenvectors.

5. Plot the points. Their coordinates are $(c_{11}, c_{21}), (c_{12}, c_{22}) \cdots (c_{1n}, c_{2n})$.

We now consider an example.

Example. A simple example is shown in Table 4.5 and Figure 4.10. The 2×5 data matrix, Data Matrix #12, is shown at the top of the table. As always, its (i, j)th element denotes the amount of species i in quadrat j, but these elements are not, in PCO, the coordinates of the data points to be ordinated. The quadrats have labeling numbers 1, ..., 5 (in italics) above the

TABLE 4.5. AN EXAMPLE OF PRINCIPAL COORDINATE ANALYSIS (SEE FIGURE 4.10).

$$\begin{matrix} & \textit{1} & \textit{2} & \textit{3} & \textit{4} & \textit{5} \end{matrix}$$

$$\mathbf{X} = \begin{pmatrix} 5 & 9 & 8 & 15 & 23 \\ 4 & 10 & 14 & 8 & 9 \end{pmatrix} \text{ is Data Matrix \#12.}$$

The matrix of squared dissimilarities is

$$\boldsymbol{\Delta} = \begin{pmatrix} 0 & 100 & 169 & 196 & 529 \\ & 0 & 25 & 64 & 225 \\ & & 0 & 169 & 400 \\ & & & 0 & 81 \\ & & & & 0 \end{pmatrix}.$$

Matrix **A** whose elements are given by Equation (4.8) is

$$\mathbf{A} = \begin{pmatrix} 120.48 & 12.48 & 12.88 & -25.92 & -119.92 \\ & 4.48 & 26.88 & -17.92 & -25.92 \\ & & 74.28 & -35.52 & -78.52 \\ & & & 23.68 & 55.68 \\ & & & & 168.68 \end{pmatrix}.$$

An eigenanalysis of **A** shows that its eigenvalues (to two decimal places) are

$$\lambda_1, \lambda_2, \lambda_3, \lambda_4, \lambda_5 = 310.77, 85.78, 4.50, 0, -9.45.$$

The first two eigenvectors are

$$\mathbf{u}_1' = (-0.517 \quad -0.152 \quad -0.3111 \quad 0.233 \quad 0.747)$$
$$\mathbf{u}_2' = (-0.629 \quad 0.169 \quad 0.721 \quad -0.234 \quad -0.027).$$

Hence the first two rows of $\mathbf{C} = \Lambda^{1/2}\mathbf{U}$, which are the required coordinates, are

$$\begin{pmatrix} \mathbf{c}_1' \\ \mathbf{c}_2' \end{pmatrix} = \begin{pmatrix} -9.117 & -2.675 & -5.485 & 4.101 & 13.176 \\ -5.823 & 1.567 & 6.681 & -2.171 & -0.254 \end{pmatrix}.$$

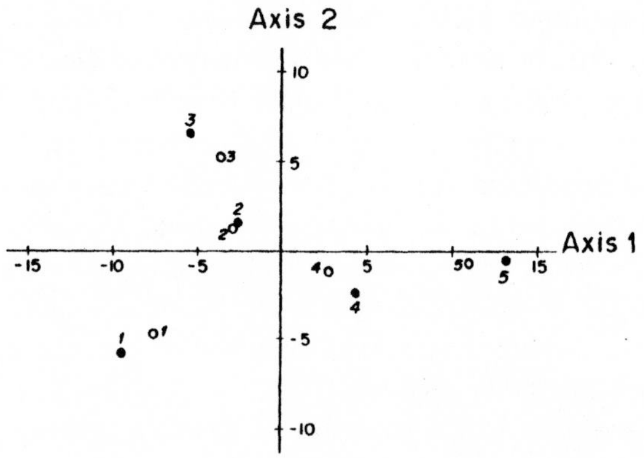

Figure 4.10. The solid dots are the data points (projected onto two-space) yielded by a PCO of data matrix **X** (Data Matrix #12) in Table 4.5. Each point is labeled with a number denoting the column of **X** that represents it. The hollow dots show the same data after unstandardized, centered PCA.

respective columns of **X**, and these are used to label the points in Figure 4.10.

It is now necessary to choose a dissimilarity measure for measuring the dissimilarity between every pair of quadrats. Let us use the city-block distance CD (page 45). Then the dissimilarity between quadrats 3 and 5, for instance, is

$$\mathrm{CD}(3,5) = |x_{13} - x_{15}| + |x_{23} - x_{25}| = |8 - 23| + |14 - 9|$$
$$= 15 + 5 = 20.$$

The dissimilarity between every pair of quadrats is measured in this way, and the squared dissimilarities are the elements of the 5×5 matrix $\boldsymbol{\Delta}$ shown in Table 4.5; all the elements on the main diagonal of $\boldsymbol{\Delta}$ are, of course, zero since $\mathrm{CD}(j, j) = 0$ for all j.

Next, the elements of **A** are determined from Equation (4.8). For example,

$$a_{35} = \tfrac{-400}{2} + \tfrac{763}{10} + \tfrac{1235}{10} - \tfrac{3916}{50} = -78.52.$$

Since matrices $\boldsymbol{\Delta}$ and **A** are symmetric, only their upper right halves are written out.

A is then analyzed. Its eigenvalues are given in the table and also its first two eigenvectors.

It will be seen that λ_5 is negative. This implies that it is impossible to arrange five points in a real space of any number of dimensions in such a way that their pairwise distances² shall have exactly the values in Δ. However, since λ_5 is much smaller in absolute magnitude than λ_1 and λ_2, the distortion introduced can safely be ignored. We are, in any case, only interested in representing the points in two-space. To use more than two dimensions in the present example (in which there are only two species) would defeat the purpose of an ordination, which is to reduce the dimensionality of the displayed data. With only two species, one can easily plot the raw data **X** in two dimensions.

The first two principal coordinates of the data points, which are the elements of the first two rows of **C**, are then calculated from

$$\begin{pmatrix} \mathbf{c}_1' \\ \mathbf{c}_2' \end{pmatrix} = \begin{pmatrix} \lambda_1^{1/2}\mathbf{u}_1' \\ \lambda_2^{1/2}\mathbf{u}_2' \end{pmatrix}.$$

These points are shown as the solid dots in Figure 4.10. For comparison, the results of carrying out an unstandardized, centered PCA on the same data are also shown (by hollow dots) on the same figure.

It is interesting to compare the desired interpoint distances (the squares of these distances are the elements of Δ) and the actual interpoint distances in the two-dimensional ordination yielded by PCO. The two matrices to be compared are shown in Table 4.6.

TABLE 4.6. A COMPARISON OF (1) INTERQUADRAT DISSIMILARITIES (CITY-BLOCK MEASURE) AND (2) INTERPOINT DISTANCES FOLLOWING THE TWO-DIMENSIONAL PCO SHOWN IN TABLE 4.5 AND FIGURE 4.10.

Interquadrat Dissimilarities[a]	Interpoint Distances[b]
$\begin{pmatrix} 0 & 10 & 13 & 14 & 23 \\ & 0 & 5 & 8 & 15 \\ & & 0 & 13 & 20 \\ & & & 0 & 9 \\ & & & & 0 \end{pmatrix}$	$\begin{pmatrix} 0 & 9.8 & 13.0 & 13.7 & 23.0 \\ & 0 & 5.8 & 7.7 & 16.0 \\ & & 0 & 13.0 & 19.9 \\ & & & 0 & 9.3 \\ & & & & 0 \end{pmatrix}$

[a] The elements of Δ in Table 4.5 are the squares of these dissimilarities.

[b] Computed, using Pythagoras's theorem, from the coordinates of the points which are given in the 2×5 matrix at the bottom of Table 4.5.

As can be seen, the discrepancies between desired and actual distances are slight. Such as they are, they result from two causes. First, it is impossible to plot the points with exactly the desired distances between every pair of them in any real space whatever, no matter how many dimensions the space has; the fact that matrix **A** has a negative eigenvalue shows that the desired configuration is impossible. Second, the "best" PCO representation of the data would be obtained by using as many axes as there are positive eigenvalues of **A**, in the present example, three (note that $\lambda_4 = 0$ and hence $\mathbf{c}_4' = \mathbf{0}$). Projecting the "best" configuration onto a space of fewer dimensions (in the present case, two dimensions) is a further cause of the discrepancies between desired and actual distances.

The problem now arises: how close an approximation between the ds (the interpoint distances) and the δs (the interquadrat dissimilarities) is "good enough"? As we have seen, if some of the eigenvalues of **A** are negative, it follows that a perfect configuration of the points (i.e., one in which $d(j, k) = \delta(j, k)$ for all (j, k) is unattainable. According to Gower and Digby (1981), "there will be difficulties if [the negative eigenvalues] dominate the positive eigenvalues." The decision as to whether the discrepancies introduced by negative eigenvalues are large enough to affect the interpretation of a PCO ordination is inevitably somewhat subjective. Unless the negative eigenvalues are of very small absolute value (say less than $\lambda_2/10$ when one is doing a two-dimensional ordination), it is probably worthwhile to compute a pair of matrices such as the pair in Table 4.6, which permit the ds and δs to be directly compared.

No research appears to have been done that might answer the question: Are some dissimilarity measures better than others if a PCO is to be done? The problem deserves investigation.

A good example of the practical application of PCO is found in Kempton (1981). The organisms studied were moths and the sampling units ("quadrats" in the terminology of this book) were light traps placed at 14 locations throughout Great Britain. Kempton was concerned with discovering whether an ordination (by PCO) of these 14 locations based on one season's moth collections remained roughly the same year after year. The question is: Is an ordination of moth communities of long-term validity, or is there so much variation from year to year that the analysis of a single year's observations is virtually meaningless? This, clearly, is an important, though rarely pondered, problem. Kempton found that there was "some consistency" among the ordinations obtained in six consecutive years, a fairly reassuring result for biogeographers who wish to ordinate geographic sites on the basis of community composition in short-lived organisms.

4.5. RECIPROCAL AVERAGING, OR CORRESPONDENCE ANALYSIS

Reciprocal averaging and correspondence analysis are alternative names for the same technique, one that is deservedly popular for ordinating ecological data. It is commonly known by the acronym RA (Gauch, 1982a). It is yet another version of PCA (besides those discussed on page 152, and many others) and, as such, might seem to have no claim to special mention. However, as we shall see, it has one great merit shared by no other version of PCA.

Thus consider one-dimensional ordination. Recall (page 83) that, by one definition, a one-dimensional ordination consists in assigning a score to each quadrat so that the quadrats can be ordered ("ordinated") along a single axis according to these scores. Each quadrat's score is the weighted sum of the species-quantities it contains. What differentiates one ordination technique from another is the system used for assigning weights to the species.

In RA the quadrats and the species are ordinated simultaneously. Scores are assigned to each quadrat and to each species in such a way as to maximize the correlation between quadrat scores and species scores (as explained later).

In the discussion, we begin by considering RA as merely another version of PCA. After that, it is shown how the same result (a scoring system for both species and quadrats) can be obtained by the so-called "reciprocal averaging" procedure.

RA as a Form of PCA

It was pointed out earlier that an "ordinary" PCA can be done in one of four different ways. One chooses first whether to center the data or leave them uncentered. Then, independently of this first choice, one chooses whether to standardize the data or leave them unstandardized (to standardize them, the elements in each row of the raw data matrix are divided by the standard deviation of all the elements in the row). In other words, the data matrix may be left untransformed or it may be centered, or standardized, or both centered and standardized. Whichever of the four possibilities is chosen, the next steps are the same: the data matrix (whether transformed or not) is postmultiplied by its transpose and the product matrix is then eigenanalyzed (for examples, see Table 4.4, page 153). Then each eigenvector consists of a list of "weights" to be attached to each species so that "scores" (which are weighted sums of species quantities) can be computed for each quadrat. The scores are the coordinates of the points in a plot of the ordination.

RA differs from the four versions of PCA already discussed in the way in which the data matrix is transformed before the eigenanalysis, and in the way in which the eigenvectors are transformed into scores after the eigenanalysis. We consider these two procedures in turn. They are demonstrated in Tables 4.7 and 4.8, which show the RA ordination of a 3×5 matrix (Data Matrix #13). The reasons for the operations will not become clear until we attain the same result by "reciprocal averaging." Here they are presented in recipe form, without explanation.

Since in RA scores are assigned both to quadrats and to species, the procedure yields an R-type and a Q-type ordination simultaneously. (Recall that an R-type analysis gives an ordination of quadrats and a Q-type analysis an ordination of species.) In the following account we first consider the Q-type part of the analysis which gives the species scores (Table 4.7), and then the R-type part of the analysis, which gives the quadrat scores (Table 4.8).

The data are not centered and they are transformed as follows. Each element in the data matrix is divided by the square root of its row total and by the square root of its column total.

As always, let the number of species (the number of rows in the data matrix **X**) be s, and the number of quadrats (the number of columns in **X**) be n.

$$\text{Let } r_i = \sum_{j=1}^{n} x_{ij} \text{ be the total of the } i\text{th row of } \mathbf{X};$$

$$\text{let } c_j = \sum_{i=1}^{s} x_{ij} \text{ be the total of the } j\text{th column of } \mathbf{X};$$

$$\text{let } N = \sum_{i=1}^{s} \sum_{j=1}^{n} x_{ij} = \sum_{j=1}^{n} c_j = \sum_{i=1}^{s} r_i \text{ be the grand total.}$$

TABLE 4.7. RA ORDINATION OF DATA MATRIX #13; THE EIGENANALYSIS GIVING THE SPECIES-SCORES.

The 3 × 5 data matrix **X** with row and column totals shown is

$$\mathbf{X} = \begin{array}{ccccc|c} 15 & 2 & 0 & 2 & 1 & 20 \\ 9 & 6 & 15 & 0 & 0 & 30 \\ 1 & 7 & 5 & 8 & 29 & 50 \\ \hline 25 & 15 & 20 & 10 & 30 & 100 \end{array}.$$

$\mathbf{M} = \mathbf{R}^{-1/2}\mathbf{X}\mathbf{C}^{-1/2}$ is

$$\begin{pmatrix} 1/\sqrt{20} & 0 & 0 \\ 0 & 1/\sqrt{30} & 0 \\ 0 & 0 & 1/\sqrt{50} \end{pmatrix} \begin{pmatrix} 15 & 2 & 0 & 2 & 1 \\ 9 & 6 & 15 & 0 & 0 \\ 1 & 7 & 5 & 8 & 29 \end{pmatrix} \begin{pmatrix} 1/\sqrt{25} & 0 & 0 & 0 & 0 \\ 0 & 1/\sqrt{15} & 0 & 0 & 0 \\ 0 & 0 & 1/\sqrt{20} & 0 & 0 \\ 0 & 0 & 0 & 1/\sqrt{10} & 0 \\ 0 & 0 & 0 & 0 & 1/\sqrt{30} \end{pmatrix}$$

$$= \begin{pmatrix} 0.67082 & 0.11547 & 0 & 0.14142 & 0.04082 \\ 0.32863 & 0.28284 & 0.61237 & 0 & 0 \\ 0.02828 & 0.25560 & 0.15811 & 0.35777 & 0.74878 \end{pmatrix}.$$

$$\mathbf{P} = \mathbf{M}\mathbf{M}' = \begin{pmatrix} 0.48500 & 0.25311 & 0.12965 \\ 0.25311 & 0.56300 & 0.17842 \\ 0.12965 & 0.17842 & 0.77980 \end{pmatrix}.$$

Eigenanalysis of **P** gives

$$\Lambda = \begin{pmatrix} 1 & 0 & 0 \\ 0 & 0.56056 & 0 \\ 0 & 0 & 0.26715 \end{pmatrix}; \quad \mathbf{U} = \begin{pmatrix} 0.44721 & 0.54772 & 0.70711 \\ 0.49281 & 0.50885 & -0.70584 \\ 0.74641 & -0.66413 & 0.04236 \end{pmatrix}.$$

The matrix of species scores is

$$\mathbf{V} = \sqrt{N}\,\mathbf{U}\mathbf{R}^{-1/2} = \begin{pmatrix} 1 & 1 & 1 \\ 1.102 & 0.929 & -0.998 \\ 1.669 & -1.213 & 0.060 \end{pmatrix}.$$

Then the (i, j)th element of the transformed matrix, say **M**, is

$$m_{ij} = \frac{x_{ij}}{\sqrt{r_i c_j}} \quad \text{with } i = 1, \ldots, s \quad \text{and } j = 1, \ldots, n.$$

The whole transformation can be compactly written in matrix form. Let **R** denote the $s \times s$ diagonal matrix whose nonzero elements are the row totals of **X**. Thus in the example in Table 4.7,

$$\mathbf{R} = \begin{pmatrix} r_1 & 0 & 0 \\ 0 & r_2 & 0 \\ 0 & 0 & r_3 \end{pmatrix} = \begin{pmatrix} 20 & 0 & 0 \\ 0 & 30 & 0 \\ 0 & 0 & 50 \end{pmatrix}.$$

Next, note that

$$\mathbf{R}^{-1/2} = \begin{pmatrix} r_1^{-1/2} & 0 & 0 \\ 0 & r_2^{-1/2} & 0 \\ 0 & 0 & r_3^{-1/2} \end{pmatrix} = \begin{pmatrix} 1/\sqrt{20} & 0 & 0 \\ 0 & 1/\sqrt{30} & 0 \\ 0 & 0 & 1/\sqrt{50} \end{pmatrix}.$$

(The reader should confirm, by matrix multiplication, that $\mathbf{R}^{-1/2}\mathbf{R}^{-1/2} = \mathbf{R}^{-1}$, and then that $\mathbf{R}^{-1}\mathbf{R} = \mathbf{R}\mathbf{R}^{-1} = \mathbf{I}$, the identity matrix.)

The $n \times n$ diagonal matrix $\mathbf{C}^{-1/2}$ is obtained from the column totals of **X** in the same way; its nonzero elements are the reciprocals of the square roots of the column totals.

If we now write

$$\mathbf{M} = \mathbf{R}^{-1/2}\mathbf{X}\mathbf{C}^{-1/2} \tag{4.9}$$

and carry out the matrix multiplication specified by the equation, it is easily seen that m_{ij}, the (i, j)th element of **M**, has the required value.

We now find the product of **M** postmultiplied by its transpose **M**′. Call the product, which is an $s \times s$ matrix, **P**. Thus

$$\mathbf{P} = \mathbf{M}\mathbf{M}'. \tag{4.10}$$

P is the matrix that must now be analyzed. The eigenanalysis is performed in the usual way, and yields an $s \times s$ diagonal matrix $\mathbf{\Lambda}$ (whose nonzero elements are the eigenvalues of **P**) and an $s \times s$ matrix **U** (whose rows are the corresponding eigenvectors of **P**). The results for the numerical example are shown in Table 4.7. It will be seen that λ_1, the largest eigenvalue of **P**, is unity. This is always the case, and the explanation is given subsequently.

It remains to derive the species scores. This is done by postmultiplying **U** by a diagonal matrix whose jth element is $\sqrt{N/r_j}$.

Denoting by **V** the $s \times s$ matrix whose rows are the sets of species scores, we therefore have, when $s = 3$,

$$\mathbf{V} = \mathbf{U}\begin{pmatrix} \sqrt{N/r_1} & 0 & 0 \\ 0 & \sqrt{N/r_2} & 0 \\ 0 & 0 & \sqrt{N/r_3} \end{pmatrix} = \sqrt{N}\,\mathbf{U}\mathbf{R}^{-1/2}. \tag{4.11}$$

(The reader should check that postmultiplying **U** by a diagonal matrix has the effect of multiplying each element in the jth column of **U** by the jth element of the diagonal matrix.)

The rows of **V** are the required vectors of species scores. It will be seen that the first row of **V**, corresponding to the largest eigenvalue, $\lambda_1 = 1$, is $(1, 1, 1)$. This result is true in general. That is, for any s, the largest eigenvalue is always $\lambda_1 = 1$. The s-element row vector of species scores corresponding to this eigenvalue, which is the first row of $\mathbf{V} = \sqrt{N}\,\mathbf{U}\mathbf{R}^{-1/2}$, is always $(1, 1, \ldots, 1)$. It is a "trivial" result, and the reason why it is invariably obtained becomes clear subsequently, when we use the reciprocal averaging procedure to do the same RA ordination. Only the rows of **V** after the first are of interest.

Now for the R-type part of the analysis, which gives the quadrat scores. Recall that for the Q-type part we analyzed the $s \times s$ matrix **P**, where

$$\mathbf{P} = \mathbf{M}\mathbf{M}' \quad \text{and} \quad \mathbf{M} = \mathbf{R}^{-1/2}\mathbf{X}\mathbf{C}^{-1/2}.$$

Notice that

$$\mathbf{M}' = \mathbf{C}^{-1/2}\mathbf{X}'\mathbf{R}^{-1/2}.$$

(It should be recalled that the transpose of the product of two matrices is the product of their respective transposes multiplied in the reverse order; see Exercise 3.9, page 131. And it should also be noticed that transposing a diagonal matrix leaves it unaltered.)

Thus, written out in full,

$$\mathbf{P} = (\mathbf{R}^{-1/2}\mathbf{X}\mathbf{C}^{-1/2})(\mathbf{C}^{-1/2}\mathbf{X}'\mathbf{R}^{-1/2}) \tag{4.12}$$

It is now clear, from considerations of symmetry, that to ordinate the quadrats we must analyze the $n \times n$ matrix

$$\mathbf{Q} = (\mathbf{C}^{-1/2}\mathbf{X}'\mathbf{R}^{-1/2})(\mathbf{R}^{-1/2}\mathbf{X}\mathbf{C}^{-1/2}). \tag{4.13}$$

Assuming that $s < n$, we find that $\mathbf{Q}$ has only s nonzero eigenvalues and they are identical with the eigenvalues of $\mathbf{P}$ (see page 127).

Table 4.8 demonstrates the analysis using Data Matrix #13. As always, $\mathbf{U}$ (which in this case is a 5×5 matrix) has the eigenvectors as its rows. The 5×5 matrix $\mathbf{W}$, whose rows are the quadrat-scores, is

$$\mathbf{W} = \sqrt{N}\,\mathbf{U}\mathbf{C}^{-1/2}; \tag{4.14}$$

this equation should be compared with (4.11).

TABLE 4.8. RA ORDINATION OF DATA MATRIX #13; THE EIGENANALYSIS GIVING THE QUADRAT-SCORES.

$\mathbf{X}'$ is the transpose of $\mathbf{X}$ in Table 4.7.
$\mathbf{M}'$ is the transpose of $\mathbf{M}$ in Table 4.7.

$$\mathbf{Q} = \mathbf{M}'\mathbf{M} = \begin{pmatrix} 0.55880 & 0.17764 & 0.20572 & 0.10499 & 0.04856 \\ & 0.15867 & 0.21362 & 0.10778 & 0.19610 \\ & & 0.40000 & 0.05657 & 0.11839 \\ & & & 0.14800 & 0.27366 \\ & & & & 0.56233 \end{pmatrix}.$$

($\mathbf{Q}$, like $\mathbf{P}$, is symmetrical and the elements below the principal diagonal have here been omitted.)

$\mathbf{Q}$ has the same nonzero eigenvalues as $\mathbf{P}$, namely,

$$\lambda_1 = 1; \qquad \lambda_2 = 0.56056; \qquad \lambda_3 = 0.26715.$$

The matrix of eigenvectors is

$$\mathbf{U} = \begin{pmatrix} 0.5000 & 0.3873 & 0.4472 & 0.3162 & 0.5477 \\ 0.6382 & 0.0273 & 0.2671 & -0.2442 & -0.6790 \\ -0.5488 & 0.1757 & 0.7739 & -0.2336 & -0.1203 \\ -0.1740 & 0.1113 & 0.0420 & 0.8657 & -0.4539 \\ -0.1061 & 0.8978 & -0.3577 & -0.1906 & -0.1359 \end{pmatrix}.$$

The matrix of quadrat scores is $\mathbf{W} = \sqrt{N}\,\mathbf{U}\mathbf{C}^{-1/2}$

$$= \begin{pmatrix} 1 & 1 & 1 & 1 & 1 \\ 1.276 & 0.070 & 0.597 & -0.772 & -1.240 \\ -1.098 & 0.454 & 1.730 & -0.739 & -0.220 \\ -0.348 & 0.287 & 0.094 & 2.738 & -0.829 \\ -0.212 & 2.318 & -0.800 & -0.603 & -0.248 \end{pmatrix}.$$

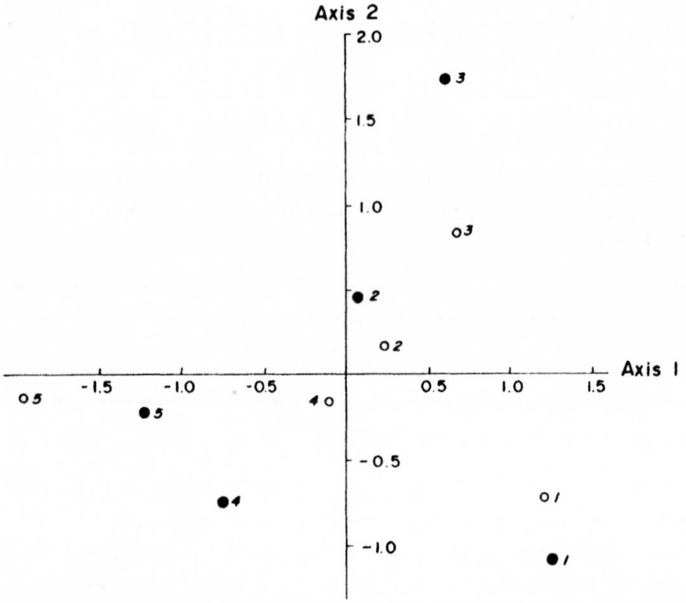

Figure 4.11. The solid dots show the outcome of RA ordination of Data Matrix #13 (Tables 4.7 and 4.8). The hollow dots show the same data after unstandardized, centered PCA; they are plotted in the plane of PCA axes 1 and 2.

As in the ordination of the species, the first set of scores (the first row of **W**) consists of ones and is of no interest. The scores on the first and second RA axes are given by the elements of the second and third rows of **W**. Using these scores as coordinates, the five points representing the quadrats give the two-dimensional RA ordination shown in Figure 4.11 (solid dots). The result of doing an unstandardized centered PCA on the same data is shown for comparison (hollow dots).

The Correlation Between Quadrat Scores and Species Scores

The analyses just described have provided sets of scores for the species (the rows of **V** in Table 4.7) and sets of scores for the quadrats (the rows of **W** in Table 4.8). Suppose the species are assigned the scores in the kth row of **V**, and the quadrats the scores in the kth row of **W**. Then, as demonstrated in Table 4.9, the square of the correlation coefficient between these scores, say

r_k^2, is equal to λ_k, the kth eigenvalue of **P** and **Q**. This holds true for $k = 1, \ldots, s$.

The correlation coefficient r_k is calculated from the formula

$$r_k = \frac{1}{N} \sum_{i=1}^{s} \sum_{j=1}^{n} x_{ij} v_{ki} w_{kj}. \tag{4.15}$$

Here x_{ij} (from the data matrix **X**) is treated as though it were "the frequency of occurrence of species i in quadrat j." Sometimes the elements of **X** are indeed frequencies; even when they are not, however, for example even when they record the biomasses of the species in the quadrats, they are treated as frequencies for the purpose of the present calculations. The term v_{ki} is the kth score of the ith species; likewise, w_{kj} is the kth score of the jth quadrat.

Table 4.9 shows the computation of r_2. As may be seen, $r_2^2 = \lambda_2$. The reader is invited to check that $r_3^2 = \lambda_3$ (see Exercise 4.8).

It is now obvious why the trivial result $\lambda_1 = 1$, with $v_1 = (1, 1, \ldots, 1)$ and $w_1 = (1, 1, \ldots, 1)$, is always obtained when **P** and **Q** are analyzed. If all the species and all the quadrats are assigned a score of unity and, equivalently, if we put $v_{1i} = 1$ and $w_{1j} = 1$ for all i and j, then the right side of (4.15)

TABLE 4.9. THE CORRELATION BETWEEN MATCHED SETS OF SPECIES SCORES AND QUADRAT SCORES FOR DATA MATRIX #13.

Computation of r_2.[a]

						Species Scores
X =	15	2	0	2	1	*1.102*
	9	6	15	0	0	*0.929*
	1	7	5	8	29	*−0.998*
Quadrat Scores	*1.276*	*0.070*	*0.597*	*−0.772*	*−1.240*	

$$r_2 = \tfrac{1}{100}\{15(1.102)(1.276) + 2(1.102)(0.070) + \cdots + 29(-0.998)(-1.240)\}$$

$$= 0.7487.$$

$$r_2^2 = 0.561 = \lambda_2.$$

[a]To the right of the data matrix is the second set of species scores (in italics), from the second row of **V** in Table 4.7. Below it (in italics) is the second set of quadrat scores, from the second row of **W** in Table 4.8.

becomes

$$r_1 = \frac{1}{N} \sum_i \sum_j x_{ij}.$$

It follows automatically that $r_1 = 1$ since, by definition, $N = \Sigma_i \Sigma_j x_{ij}$. To repeat, this trivial result is discarded when an RA ordination is done.

Another point worth repeating is the following. The species scores and quadrat scores found by RA are such as to maximize the correlation between them. The proof is beyond the scope of this book; it may be found in Anderberg (1973, p. 215).

The Reciprocal Averaging Technique

RA ordination can also be done by "reciprocal averaging." In outline the procedure is as follows. First, arbitrary trial values are chosen for the species scores. Next, a first set of quadrat scores is computed from these species scores. Then a second set of species scores is computed from the first set of quadrat scores, then a second set of quadrat scores from the second set of species scores. And so on, back and forth reciprocally, until the vectors of scores maintain constant relative proportions.

Table 4.10 illustrates the procedure numerically, using Data Matrix #13 again. At every stage, each quadrat score is the weighted average of the last-derived species scores (and vice versa). In computing these averages, the species scores being averaged are weighted by the amounts of the species in the quadrat (and *mutatis mutandis* when quadrat scores are averaged).

In symbols let $\mathbf{v}^{(0)}, \mathbf{v}^{(1)}, \ldots$ denote the successive vectors of species scores, and let $\mathbf{w}^{(0)}, \mathbf{w}^{(1)}, \ldots$ denote the successive vectors of quadrat scores. First, values for the elements of $\mathbf{v}^{(0)}$ are chosen. It is convenient to use percentages, ranging from 0 for the lowest score to 100 for the highest. In the example the chosen scores are

$$\mathbf{v}^{(0)} = \left(v_1^{(0)}, v_2^{(0)}, v_3^{(0)}\right) = (100, 50, 0).$$

The elements of $\mathbf{w}^{(0)}$ are now computed. For instance, the jth element of $\mathbf{w}^{(0)}$ is

$$w_j^{(0)} = \left[x_{1j}v_1^{(0)} + x_{2j}v_2^{(0)} + \cdots + x_{sj}v_s^{(0)}\right] / c_j \qquad (4.16)$$

where c_j is the jth column total of the data matrix.

TABLE 4.10. OBTAINING SPECIES SCORES AND QUADRAT SCORES FOR DATA MATRIX #13 BY RECIPROCAL AVERAGING.[a]

						$\mathbf{v}^{(0)}$	$\mathbf{v}^{(1)}$(%)	$\mathbf{v}^{(2)}$(%)		$\mathbf{v}$ (%)
	15	2	0	2	1	*100*	*64.0* (100)	*69.9* (100)	...	*76.9* (100)
	9	6	15	0	0	*50*	*48.8* (68.9)	*59.5* (80.1)	...	*72.3* (91.8)
	1	7	5	8	29	*0*	*15.1* (0)	*17.7* (0)	...	*20.9* (0)
$\mathbf{w}^{(0)}$	*78.0*	*33.3*	*37.5*	*20.0*	*3.3*					
$\mathbf{w}^{(1)}$	*84.8*	*40.9*	*51.7*	*20.0*	*3.3*					
$\mathbf{w}^{(2)}$	*88.8*	*45.3*	*60.1*	*20.0*	*3.3*					
.....	.....	.....	.....	.....	.....					
$\mathbf{w}$	*93.0*	*50.0*	*68.8*	*20.0*	*3.3*					
(%)	(100)	(52.1)	(73.0)	(18.6)	(0)					

V	0				91.8	100
Row 2 of **V**	− 0.998				0.929	1.102
W	0	18.6	52.1	73.0		100
Row 2 of **W**	− 1.240	−0.772	0.070	0.597		1.276

[a]The data matrix is shown above and to the left of the double line. Successive approximations to the species scores are in the columns on the right, labeled $\mathbf{v}^{(0)}, \mathbf{v}^{(1)}, \ldots$. Successive approximations to the quadrat-scores are in the rows below, labeled $\mathbf{w}^{(0)}, \mathbf{w}^{(1)}, \ldots$.

Thus in the numerical example

$$w_1^{(0)} = [(15 \times 100) + (9 \times 50) + (1 \times 0)]/25 = 78.0.$$

When the n elements of $\mathbf{w}^{(0)}$ have been found, $\mathbf{v}^{(1)}$ is computed. Its ith element $v_i^{(1)}$ is

$$v_i^{(1)} = \left[x_{i1}w_1^{(0)} + x_{i2}w_2^{(0)} + \cdots + x_{in}w_n^{(0)}\right]/r_i, \tag{4.17}$$

where r_i is the ith row total of the data matrix.

Thus in the numerical example

$$v_1^{(1)} = [(15 \times 78.0) + (2 \times 33.3) + (0 \times 37.5)$$

$$+ (2 \times 20.0) + (1 \times 3.3)]/20 = 64.0.$$

The elements of $\mathbf{v}^{(1)}$ are rescaled to percentages before they are used to compute $\mathbf{w}^{(1)}$. (The successive $\mathbf{w}$ vectors could also be rescaled but it is not necessary.)

The procedure is continued until the vectors stabilize (i.e., until any further steps give unchanged results). The final results in the example are shown by the column on the extreme right in Table 4.10 (which gives the final species scores as percentages) and the row at the bottom (which gives the final quadrat scores as percentages). As is shown in the lower part of the table, these scores are the same (apart from being rescaled as percentages) as row 2 of **V** (in Table 4.7) and row 2 of **W** (in Table 4.8). Thus they are the required scores for, respectively, a one-dimensional ordination of the species, and a one-dimensional ordination of the quadrats.

The scores on the second RA axes (i.e., the third row of **V** and the third row of **W**) can be obtained by a similar, though computationally more laborious procedure. It is not described here. Details are given by Hill (1973). The reader should confirm that if $\mathbf{v}^{(0)} = (1,1,1)$, then $\mathbf{w}^{(0)} = (1,1,1,1,1)$. This is the trivial result mentioned on page 181.

We now show the equivalence between the reciprocal averaging procedure just described and the outcomes of the eigenanalyses of matrices **P** and **Q** in Equations (4.10) and (4.13).

Suppose reciprocal averaging has been continued back and forth (rescaling the species scores as percentages each time) until stability has been reached. Then we can rewrite Equations (4.17) and (4.16) (in that order), dropping the superscripts in parentheses. Thus (4.17) becomes

$$v_i = (x_{i1}w_1 + x_{i2}w_2 + \cdots + x_{in}w_n)/r_i \qquad \text{for } i = 1,\ldots,s; \qquad (4.18)$$

(4.16) becomes

$$w_j = (x_{1j}v_1 + x_{2j}v_2 + \cdots + x_{sj}v_s)/c_j \qquad \text{for } j = 1,\ldots,n. \qquad (4.19)$$

Next let us write these two equations more compactly in matrix form.

Let $\mathbf{R}^{-1}$ be the $s \times s$ diagonal matrix whose ith element is $1/r_i$; let $\mathbf{C}^{-1}$ be the $n \times n$ diagonal matrix whose jth element is $1/c_j$. The matrix versions of (4.18) and (4.19), with the size off each matrix shown below it, are

$$\underset{(s\times 1)}{\mathbf{v}} = \underset{(s\times s)}{\mathbf{R}^{-1}} \underset{(s\times n)}{\mathbf{X}} \underset{(n\times 1)}{\mathbf{w}} \qquad (4.20)$$

and

$$\underset{(n\times 1)}{\mathbf{w}} = \underset{(n\times n)}{\mathbf{C}^{-1}} \underset{(n\times s)}{\mathbf{X}'} \underset{(s\times 1)}{\mathbf{v}}\,. \qquad (4.21)$$

Substituting the right side of (4.21) for the $\mathbf{w}$ in (4.20) gives

$$\mathbf{v} = (\mathbf{R}^{-1}\mathbf{X})(\mathbf{C}^{-1}\mathbf{X}'\mathbf{v}) \tag{4.22}$$

We now operate on (4.22). Notice that matrices may be factored or multiplied as may be convenient, provided their order is never changed. Parentheses are put in wherever they help to make the steps clearer. The reader should check the sizes of the matrices at every step to be sure that all multiplications are possible.

First premultiply both sides of (4.22) by $\mathbf{R}^{1/2}$. Then

$$\mathbf{R}^{1/2}\mathbf{v} = (\mathbf{R}^{1/2}\mathbf{R}^{-1})(\mathbf{X}\mathbf{C}^{-1}\mathbf{X}')\mathbf{v} \tag{4.23}$$

$$= \mathbf{R}^{-1/2}(\mathbf{X}\mathbf{C}^{-1}\mathbf{X}')(\mathbf{R}^{-1/2}\mathbf{R}^{1/2})\mathbf{v}. \tag{4.24}$$

Here the interpolated factor $(\mathbf{R}^{-1/2}\mathbf{R}^{1/2})$ is simply a factored form of the identity matrix and leaves the right side of the equation unchanged. The reason for interpolating it becomes clear in a moment.

Writing $\mathbf{C}^{-1} = \mathbf{C}^{-1/2}\mathbf{C}^{-1/2}$ and rearranging parentheses, we now see that

$$\mathbf{R}^{1/2}\mathbf{v} = (\mathbf{R}^{-1/2}\mathbf{X}\mathbf{C}^{-1/2})(\mathbf{C}^{-1/2}\mathbf{X}'\mathbf{R}^{-1/2})(\mathbf{R}^{1/2}\mathbf{v}). \tag{4.25}$$

On substituting from (4.12), this becomes

$$(\mathbf{R}^{1/2}\mathbf{v}) = \mathbf{P}(\mathbf{R}^{1/2}\mathbf{v}). \tag{4.26}$$

Both sides of (4.26) are $s \times 1$ matrices (i.e., s-element column vectors). Now transpose both sides to convert them to row vectors. Thus transposing the left side gives

$$(\mathbf{R}^{1/2}\mathbf{v})' = \mathbf{v}'\mathbf{R}^{1/2},$$

and transposing the right side gives

$$\left[\mathbf{P}(\mathbf{R}^{1/2}\mathbf{v})\right]' = (\mathbf{R}^{1/2}\mathbf{v})'\mathbf{P}' = \mathbf{v}'\mathbf{R}^{1/2}\mathbf{P}' = \mathbf{v}'\mathbf{R}^{1/2}\mathbf{P}.$$

The last equality follows from the fact that $\mathbf{P}$ is symmetrical so that $\mathbf{P} = \mathbf{P}'$. Hence

$$(\mathbf{v}'\mathbf{R}^{1/2}) = (\mathbf{v}'\mathbf{R}^{1/2})\mathbf{P}. \tag{4.27}$$

It follows that $(\mathbf{v}'\mathbf{R}^{1/2})$ is an eigenvector of $\mathbf{P}$. This result is true for all s

vectors of species scores, that is, for all s rows of $\mathbf{V}$. Hence

$$\mathbf{V}\mathbf{R}^{1/2} \propto \mathbf{U} \tag{4.28}$$

where $\mathbf{U}$ is the $s \times s$ matrix whose rows are the eigenvectors of $\mathbf{P}$. Postmultiplying both sides of (4.28) by $\mathbf{R}^{-1/2}$ shows that

$$\mathbf{V} \propto \mathbf{U}\mathbf{R}^{-1/2} \tag{4.29}$$

which, apart from the constant of proportionality $\sqrt{N}$, is identical with (4.11).

This explains why the species scores can be obtained either by reciprocal averaging or by eigenanalysis of $\mathbf{P}$; the results are the same. It is left to the reader (Exercise 4.9) to derive the analogous relation between matrix $\mathbf{Q}$ in Equation (4.13) and the vectors of quadrat scores.

4.6. LINEAR AND NONLINEAR DATA STRUCTURES

The methods of ordination discussed in this chapter so far (PCA, PCO, and RA) are all achieved by projecting an s-dimensional swarm of data points onto a space of fewer dimensions. In the simplest method (PCA) the coordinates of the points before projection are the measured quantities of the s species in each of the n quadrats; centering and standardizing the data (both optional) merely amount to changing the origin and the scale of measurement, respectively. In PCO and RA, the measurements are adjusted in a more elaborate fashion (as described in Sections 4.4 and 4.5) before the swarm is projected onto a space of fewer than s dimensions. But, to repeat, the final step in all these ordinations consists in projecting a swarm of points onto a line, a plane, or a three-space.

It is obvious that whenever such a projection is done, there is a risk that the original pattern of the swarm will be misinterpreted; this risk is the price that must be paid for a reduction in dimensionality. We now ask whether projection of the swarm is likely to produce a pattern that is positively misleading. The answer depends on whether the original data swarm has a *linear* or *nonlinear structure*.

Figure 4.12 demonstrates the difference. The three-dimensional swarm in the upper panel has a linear structure; if a one or two-dimensional ordination of the swarm were done by projecting the points onto a line or plane,

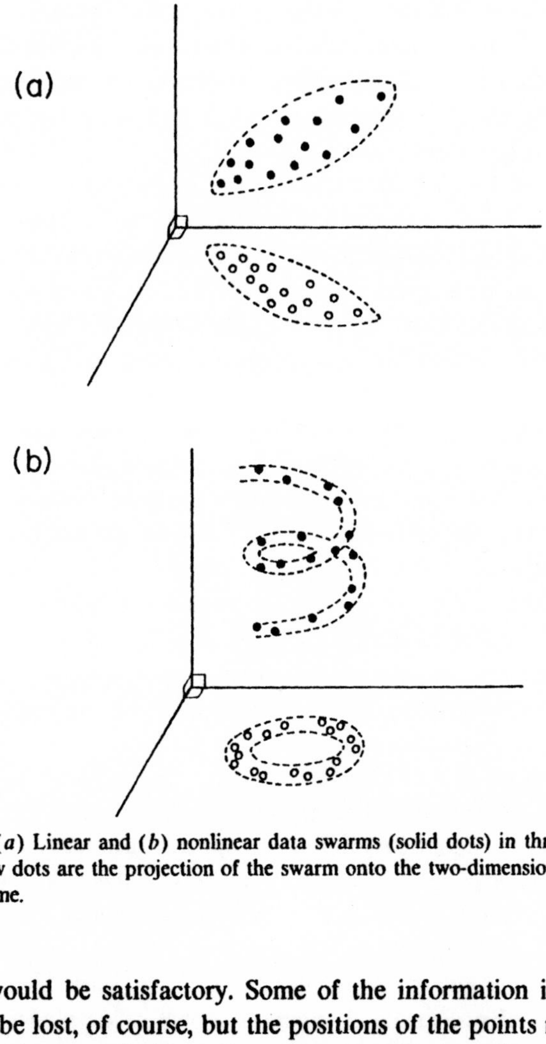

Figure 4.12. (*a*) Linear and (*b*) nonlinear data swarms (solid dots) in three-space. In each case the hollow dots are the projection of the swarm onto the two-dimensional "floor" of the coordinate frame.

the result would be satisfactory. Some of the information in the original data would be lost, of course, but the positions of the points relative to one another would be reasonably well preserved in the sense that points close to each other in the original three-dimensional swarm would remain close to each other in the one or two-dimensional projections.

The spiral swarm in the lower panel has a nonlinear structure. There is obviously no way of orienting a line or a plane so that when the swarm is projected onto it the relationships of all the points to one another are even approximately preserved. For instance, suppose the swarm were projected onto the floor of the coordinate frame; it would be found that the points at each end of the spiral, which are far apart in three-space, would be close together in two-space. Indeed, if the two-dimensional picture were the only available representation of the swarm, it would be impossible to judge whether its original three-dimensional shape had been that of a spiral, a hollow cylinder, or a doughnut.

It should now be clear that ordination by projection, for example by PCA, PCO, or RA, although entirely satisfactory if the data swarm is linear, may give misleading results if the swarm is nonlinear. It is sometimes said that PCA, for example, gives a distorted representation of nonlinear data. This is a misuse of the word "distorted." The picture of a many-dimensional swarm that PCA yields is no more distorted than, say, a photograph in which both distant and nearby objects appear. One would not call such a picture distorted because the images of a distant mountain peak and a nearby tree-top, say, were close together on the paper. In the same way, the circle of points on the floor of the coordinate frame in Figure 4.12*b* is not in the least distorted. But it is misleading. What we require is a method of ordination that deliberately introduces distortion of a well-planned, specially designed kind, that will correct the misleading impression sometimes given by truly undistorted data.

Various methods of ordination that achieve this result have been devised. They are known collectively as nonlinear ordination methods. A note on terminology is necessary here. The contrast between linear and nonlinear ordination methods is that they are appropriate for linear and nonlinear data structures, respectively. The term "linear ordination" should not be used (though it occasionally is) to mean a one-dimensional, as opposed to a two or three-dimensional, ordination. The term *catenation*, suggested by Noy-Meir (1974), is a useful and unambiguous synonym for "nonlinear ordination."

We now consider how nonlinear data swarms can arise in practice. Then a good method of ordinating such data, known as detrended correspondence analysis, is described.

The Arch Effect

Ecological data often have a nonlinear structure. An example of an investigation that would yield such data is worth considering in detail.

Imagine an ecological community occupying a long environmental gradient, for instance the vegetation on a mountainside. The vegetation forms a coenocline, a community whose species-composition changes smoothly and continuously along the gradient in response to continuous change in environmental conditions. If the species-composition is examined along an ascending transect, from the foot of the mountain to its summit, one finds that the valley-bottom species gradually dwindle in abundance, each at its own rate. They are replaced, as one ascends the mountain, by a sequence of new species, each of which appears, becomes gradually more abundant, attains its maximum abundance, and then gradually diminishes in abundance again. If the mountain is high, the turnover of species is likely to be complete. None of those growing at the bottom will be found at the top.

According to the Gaussian model of community structure (Gauch and Whittaker, 1972; and see Gauch, 1982a), the species all respond independently of one another to the gradient. The response of any one species can be represented by a Gaussian (i.e., normal, bell-shaped) curve. Each species' response curve, therefore, has three properties: the location of its peak (the altitude at which the species grows best), the height of its peak (the maximum density of the species), and its dispersion or spread (the width of the altitude zone it occupies). Every species has its own individual values of these properties.

Figure 4.13*a* gives a diagrammatic portrayal of such a coenocline. Each curve represents one species, the horizontal axis measures distance along the gradient, and the height of a particular species' curve above this axis shows the way the species responds to the varying environmental conditions along the gradient.

Now imagine that the coenocline is sampled by placing a row of quadrats spaced at equal intervals along the gradient (up the mountainside). Will the resultant "data structure" (the shape of the swarm of data points) be linear or nonlinear? The answer depends on the length of the segment of gradient that is sampled.

Suppose the sampled segment is long, and contains the peaks of one or more species' response curves. These species do not respond monotonically to the gradient over the length of the segment; that is, they do not increase continuously, or decrease continuously, along it. Rather, they first increase and then decrease. As a consequence, the data structure is nonlinear.

But if sampling is confined to only a short segment of the gradient (Figure 4.13*b*) over which the response curve of each species present in the segment is at least approximately linear, then the data structure itself is (again, approximately) linear.

We now consider the shape of the nonlinear data swarm that results when several species first increase, and then decrease, along a sampled

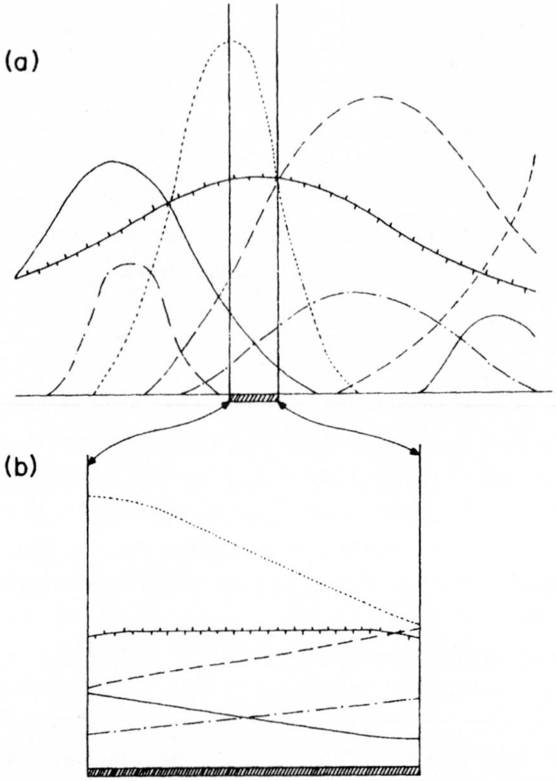

Figure 4.13. (a) The response curves of eight species along an environmental gradient. If the community (a coenocline) were sampled at a series of points along its whole length, the data swarm would be nonlinear. (b) An enlarged segment of (a). If the coenocline were sampled at a number of closely spaced points within the segment, which is so short that the response curves are not appreciably nonlinear within it, the data swarm would be linear (at least approximately).

gradient. Figure 14.4a shows a very simple artificial example; there are only three species and their response curves are identically shaped and equally spaced. Assume that $n = 12$ quadrats are examined; they are located at the sites marked on the horizontal axis. The reader is invited to visualize (or construct, with stiff wire) the curve in three-space connecting the sequence of data points these quadrats would yield (each point has as coordinates the amounts of each of the three species in the quadrat that the point represents). It will be found that the curve is the same as that shown by the

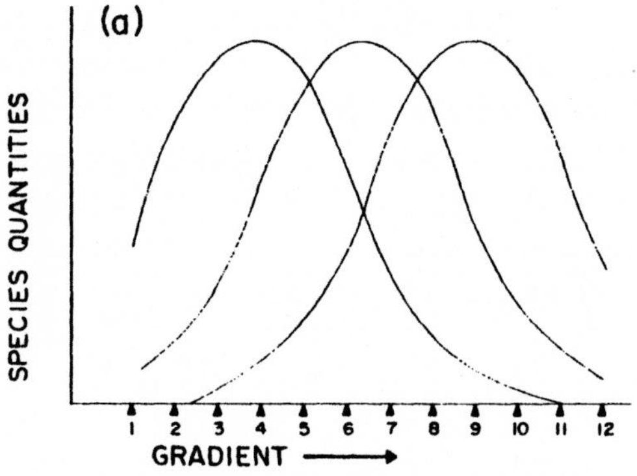

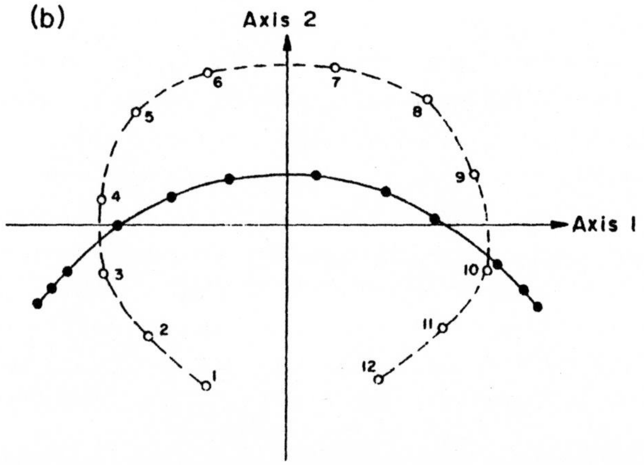

Figure 4.14. (*a*) The response curves of 3 species along a gradient; 12 quadrats are located at the numbered points marked with arrowheads (artificial data). (*b*) Ordinations of the 12 data points by PCA (hollow dots, dashed line) and by RA (solid dots, solid line). Both ordinations exhibit the arch effect. The RA ordination also shows scale contraction at both extremities.

dashed line in Figure 4.14*b*. The latter is a two-dimensional PCA ordination of the data, and as such is an undistorted "picture" of the points; it shows the pattern produced when they are projected onto the plane that fits them most closely. The curve exhibits the so-called *arch effect* (sometimes called the *horseshoe effect*), and the fact that it appears when data from a long gradient are ordinated by PCA (and also, as shown later, by RA) detracts from the usefulness of PCA as an ordination method. The drawback is this:

one would like the result of ordinating the quadrats observed along a linear gradient to have a linear pattern themselves, in the present case to form a more or less straight row in two-space. But they do not. The dashed curve in Figure 4.14*b*, which is almost a closed loop, gives a misleading idea of the gradient even though it is an undistorted picture of the data swarm. Suppose one were to ask for a one-dimensional ordination of these data. The ordering of the quadrats along axis 1 turns out to be

$$3 \quad 4 \quad 5 \quad 2 \quad 1 \quad 6 \quad 7 \quad 12 \quad 8 \quad 11 \quad 9 \quad 10,$$

an obviously meaningless result. But if ordination by PCA (or by RA or by PCO) is performed on data with a linear structure (as in Figure 4.13*b*), the result accords with what one intuitively expects; for an example, see Exercise 4.10.

It might be argued that the PCA ordination in Figure 4.14*b* would not mislead in practice. The points are numbered according to their position on the gradient and can be joined, in proper order, by a smooth curve. But it should be recalled that this is an artificial example with only three species. Given field data with many species, one always has to project the data swarm onto a space of far fewer dimensions than it occupied originally; if the swarm is a "hyper-coil" (a multidimensional analogue of the dashed curve in Figure 4.14*b*), then when it is projected it will automatically "collapse" and yield as meaningless a pattern in the line, the plane, or three-space as the one-dimensional ordering of the artificial example listed previously.

The problem is, of course, compounded when the gradient sampled is less obvious than that of a mountainside, or is not even apparent at all. Indeed, if environmental variables such as soil moisture, soil texture, and the like are varying haphazardly in space, there may be no gradient in the ordinary sense. Then the quadrats will have no particular ordering before an analysis is carried out, and the purpose of the analysis is to perceive their ordering (if there is any) and diagnose its cause.

What is required, therefore, is an ordination method that is not subject to the arch effect, one which will ordinate a nonlinear data swarm in a way that clearly exhibits in one, two, or three dimensions the true interrelationships among the quadrats.

Let us first consider whether RA is an improvement over PCA. The solid curve in Figure 4.14*b* shows the RA ordination of the same 12 quadrats and represents the sort of results that RA is found to yield in practice with real data. It is an improvement over PCA in that the arch effect is less exaggerated. However, the effect is still present in mild form, and although it does not give an erroneous ordering on axis 1, it still produces a meaningless pattern in the direction of axis 2; a true representation of the 12 quadrats would obviously consist of a row of equispaced points along axis 1 with no component on axis 2. Also, there is a contraction of scale at each end of the gradient: the points at the end are more closely spaced than those at the middle and this variation in spacing does not correspond to any variation in the steepness of the environmental gradient.

Therefore, although RA does better than PCA in giving a true representation of the interrelationships of the quadrats, there is room for further improvement. One way of achieving this is to use detrended correspondence analysis, an ordination method that we now consider.

Detrended Correspondence Analysis

Detrended correspondence analysis (DCA) is an ordination method that overcomes the two defects of ordinary RA. It flattens out the misleading arch, and it corrects the contraction in scale at each end of an RA-ordinated data swarm. DCA does this by applying the requisite adjustments to an ordinary RA ordination. In general terms, the adjustments (to a two-dimensional ordination) are as follows.

The arch effect is removed by dividing the RA-ordinated data swarm into several short segments with dividing lines perpendicular to the first axis, and then sliding the segments with respect to one another so that the arch disappears. More precisely, the segments are shifted up or down in such a way that the average height of the points within each segment (i.e., the average of their scores on the second axis) are all equal. The scale contraction at each end of the swarm is corrected by straightforward rescaling of those parts of the axes where it is needed. A detailed description of these procedures and a FORTRAN program for carrying them out have been given by Hill (1979a), who devised the method. The procedures consist in overt, systematic data manipulation, carried out in order to force an ordination into a form that accords as well as possible with intuitive expectations. Therefore, there is a risk that erroneous intuitions may sometimes be forced upon a body of data. However, experience with real data and tests with artificial data (see Gauch, 1982a and references therein) suggest that the method gives useful results and permits ecologically correct interpretations to be derived from confusing multivariate data.

An example, using real data, of the contrast between RA and DCA is shown in Figure 4.15. The data consist of observations on the aquatic vegetation at 37 sites in oxbow lakes in the floodplain of the Athabasca River in northern Alberta. The lakes differed among themselves in a variety of abiotic factors, the most important of which was salinity. Ordination of

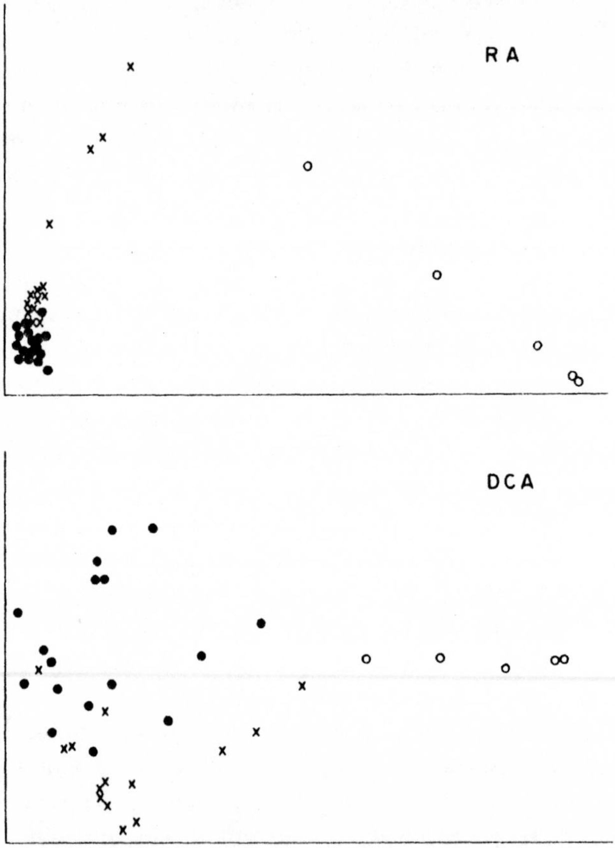

Figure 4.15. RA and DCA ordinations of 37 oxbow lakes in the valley of the Athabasca River ordinated on the basis of the communities of aquatic angiosperms (plus *Chara* and *Nitella*) growing in them. Three classes of site are distinguished: those with *Typha latifolia* (●), those with *Triglochin maritima* (○), and those with neither species (×). [The DCA ordination is adapted from Lieffers (1984). The RA ordination of the same data was kindly provided by V.J. Lieffers (pers. comm.).]

the sample sites by RA and by DCA are shown in the upper and lower panels of the figure. Different symbols have been used for the sites depending on whether they contained *Typha latifolia* (which cannot tolerate saline water), *Triglochin maritima* (which thrives in saline water), or neither; the two species never occurred together. The RA ordination clearly exhibits both the arch effect and the scale contraction effect, and both effects disappear when the data are ordinated by DCA.

4.7. COMPARISONS AND CONCLUSIONS

All four of the ordination methods described in this chapter have merit in appropriate circumstances.

Three of the methods are suitable for data with a linear structure, and such data are obtained very frequently in ecological studies (van der Maarel, 1980). PCA has the merit that it is the most straightforward, conceptually, of all methods; it allows the user to look at a visible projection of a multidimensional and hence unvisualizable swarm of points. In addition, uncentered PCA aids in the recognition of distinct classes of quadrats (see page 162). PCO sometimes allows one to construct and inspect a data swarm in which the distances between every pair of points corresponds (approximately) with some chosen measure of their dissimilarity. RA provides simultaneous ordinations of quadrats and species.

With nonlinear data, DCA is the ordination method at present favored by the majority of ecologists. Its advantages are that it removes two likely sources of error that arise when nonlinear data are ordinated by ordinary RA, namely, the arch effect and the scale contraction effect. Its defect is that it gains these advantages by data manipulation, that is, by deliberately flattening the arch and by applying local adjustments to the scales of the axes. If these troublesome effects are truly mathematical artifacts devoid of ecological meaning, then it is obviously desirable to remove them. But overzealous correction of suspected "defects" may sometimes lead to the unwitting destruction of ecologically meaningful information.

Other methods of ordinating nonlinear data exist but are beyond the scope of this book. An especially promising method has been devised by Shepard and Carroll (1966); it is also described in Noy-Meir (1974), and briefly in Pielou (1977). It is known as "parametric mapping" or, alternatively, as "continuity analysis." It is mathematically more difficult than

DCA, but is free of the rather contrived "corrections" that make DCA ordinations somewhat subjective. Parametric mapping could profitably be tested in ecological contexts; it may prove to have the merits of DCA without its defect of artificiality.

There is, however, a valuable byproduct of RA and DCA ordinations that more than compensates for any defects they may have. They provide the information needed to rearrange the rows and columns of a data matrix in such a way as to make the raw data themselves easily interpretable. This is possible because both methods ordinate the quadrats and the species simultaneously.

Consider the artificial example in Table 4.11. (An artificial example is used for the sake of clarity.) The two matrices in the table contain identical information. Both record the abundances of 10 species in 10 quadrats. The upper matrix shows the data as they might have been collected in the field. It displays no discernible pattern, and there is no reason to suspect that it contains a concealed pattern. It is typical of the sort of matrices obtained when observations are first recorded in a field notebook. Neither the species nor the quadrats are listed in any particular order; indeed, one often does not know what the intrinsic ordering is, or even if there is any.

Now suppose that these data are ordinated by RA or DCA. We require only a one-dimensional ordination from which the order of the points along the axis can be obtained. (Since RA and DCA range the points in identical order on the first axis, either method may be used.) The quadrats and the species are then reordered according to the magnitudes of their scores. In the example, the ordering of the quadrats on the first axis is

$$7, 1, 10, 9, 5, 2, 8, 6, 4, 3,$$

and the ordering of the species is:

$$2, 9, 8, 1, 4, 6, 10, 7, 5, 3.$$

Let the data matrix be rewritten with the quadrats (columns) arranged in the order shown in the first list, and the species (rows) arranged in the order shown in the second list. The result is the lower matrix in Table 4.11. The pattern or "structure" of the data is now strikingly obvious.

An example using real data is given by Gauch (1982a).

An alternative method of rearranging data matrices in order to exhibit their structure has been devised by van der Maarel, Janssen, and Louppen (1978), who give a program, TABORD, for carrying it out.

TABLE 4.11. REARRANGING A DATA MATRIX TO REVEAL ITS INTRINSIC STRUCTURE.

The raw, unordered data matrix

Species \ Quadrat	*1*	*2*	*3*	*4*	*5*	*6*	*7*	*8*	*9*	*10*
1	2	2	—	—	3	—	1	1	4	3
2	3	—	—	—	—	—	4	—	1	2
3	—	—	4	3	—	2	—	1	—	—
4	1	3	—	—	4	1	—	2	3	2
5	—	1	3	4	—	3	—	2	—	—
6	—	4	—	1	3	2	—	3	2	1
7	—	2	2	3	1	4	—	3	—	—
8	3	1	—	—	2	—	2	—	3	4
9	4	—	—	—	1	—	3	—	2	3
10	—	3	1	2	2	3	—	4	1	—

The same data with the rows and columns rearranged

Species \ Quadrat	*7*	*1*	*10*	*9*	*5*	*2*	*8*	*6*	*4*	*3*
2	4	3	2	1	—	—	—	—	—	—
9	3	4	3	2	1	—	—	—	—	—
8	2	3	4	3	2	1	—	—	—	—
1	1	2	3	4	3	2	1	—	—	—
4	—	1	2	3	4	3	2	1	—	—
6	—	—	1	2	3	4	3	2	1	—
10	—	—	—	1	2	3	4	3	2	1
7	—	—	—	—	1	2	3	4	3	2
5	—	—	—	—	—	1	2	3	4	3
3	—	—	—	—	—	—	1	2	3	4

EXERCISES

4.1. Consider Table 4.2. What are the eigenvalues of the covariance matrix yielded by the SSCP matrix **R**?

4.2. Let **X** be a row-centered data matrix and let **Y** = **UX** be the transformed matrix obtained by doing a PCA. What do the quantities tr(XX′) and tr(YY′) represent in geometric terms? Why would you expect them to be equal? [Reminder: tr(A) is the *trace* of matrix A, i.e., the sum of the elements on the main diagonal.]

4.3. Show that the eigenvectors of a 2 × 2 correlation matrix are always

$$(0.7071 \quad 0.7071) \text{ and } (-0.7071 \quad 0.7071).$$

4.4. Refer to Table 4.4 and Figure 4.6. From the table, determine the angles between: (a) the x_1-axis and the y_1-axis in Figure 4.6*a*; (b) the x_1-axis and the y_1'-axis in Figure 4.6*a*; (c) the (x_1/σ_1)-axis and the y_1''-axis in Figure 4.6*b*; (d) the (x_1/σ_1)-axis and the y_1'''-axis in Figure 4.6*b*.

4.5. Refer to page 164. What is the coefficient of asymmetry of axis 3 in the example described in the text? (Note: axis 3 does not appear in Figure 4.9*b* because it is perpendicular to the plane of the page.)

4.6. Let A, M, N, and I all be $n \times n$ matrices. A is the matrix whose (i, j)th element is given in paragraph 12, page 170. The (i, j)th element of M is $-\frac{1}{2}\delta^2(j, k)$. All the elements of N are equal to $1/n$. I is the identity matrix. Show that (I − N)M(I − N) = A.

4.7. The quantities of two species in quadrats A, B, and C are given by the data matrix

$$\mathbf{X} = \begin{pmatrix} 1 & 5 & 5 \\ 4 & 4 & 7 \end{pmatrix} \quad \text{(columns } A, B, C\text{)}.$$

Perform a PCO on these data by simple geometric construction, using city-block distance as measure of the dissimilarity between quadrats. (Show the result as a diagram of the pattern of the three points after the ordination; *do not* compute the coordinates of the points.)

4.8. Refer to Tables 4.7, 4.8, and 4.9. Confirm that

$$r_3^2 = \lambda_3.$$

Here r_3 is the correlation between the species scores in row 3 of V (Table 4.7) and the quadrat scores in row 3 of W (Table 4.8).

4.9. Refer to Equation (4.27) on page 187, showing how the species scores for a RA ordination are related to the eigenvectors of **P** defined in Equation (4.12) (page 181). Derive the analogous relation between the quadrat scores and the eigenvectors of **Q** defined in Equation (4.13).

4.10. Consider the following data matrix in which the rows represent species and the columns quadrats.

$$\mathbf{X} = \begin{pmatrix} 11 & 12 & 13 & 14 \\ 17 & 19 & 21 & 23 \\ 22 & 27 & 32 & 37 \\ 30 & 27 & 24 & 21 \\ 34 & 28 & 22 & 16 \end{pmatrix}.$$

Find matrix **Y**, giving the coordinates of the four points after an unstandardized, centered PCA of the data. (Hint: with these data there is no need to construct the covariance matrix and do an eigenanalysis. To perceive the structure of the data, inspect the results of plotting against each other the quantities of every pair of species.)

Chapter Five

Divisive Classification

5.1. INTRODUCTION

In this chapter we return to the topic of classifying ecological data. Several methods of classification were described in Chapter 2; all were so-called agglomerative methods. Here we consider divisive methods. The distinction is as follows.

In an *agglomerative classification* one begins by treating all the quadrats (or other sampling units) as separate entities. They are then combined and recombined to form successively more inclusive classes. The process is often called "clustering." Metaphorically, construction of the classificatory dendrogram (tree diagram) starts with the twigs and progresses towards the trunk.

A *divisive classification* goes the other way. It starts with the trunk and progresses towards the twigs. That is, the whole collection of quadrats is treated as a single entity at the outset. It is then divided and the subdivisions redivided, again and again.

Compared with divisive methods, agglomerative methods have one notable disadvantage. It arises because they start with the smallest units (the quadrats themselves). If, by chance, there are a few atypical quadrats in the data set, these quadrats are likely to have a strong effect on the first round of a clustering process, and "bad" fusions at the beginning will influence all later fusions. The obvious (but, with one exception, impracticable) solution is to adapt the agglomerative methods so that they can be used the other way round. It is easy (in theory) to devise a method of classification by division that proceeds as follows. The whole collection of quadrats is first divided into two groups in every conceivable way, and one then judges which of the ways is "best" according to some chosen criterion. If there are n quadrats, there will be $2^{n-1} - 1$ different divisions to compare with one another (for a proof, see Pielou, 1977). Having discovered the best possible division to make at the first stage, the whole process must be repeated on each of the two classes identified at this stage, and then on each of the four classes identified at the second stage, then on each of the eight classes identified at the third stage, and so on. Not surprisingly, the computer requirements of such methods are so excessive that they are infeasible unless n is very small.

However, there is another method (actually, a whole set of related methods) of doing divisive classifications that avoids these computational difficulties. It consists in first doing an ordination of the data (in any way one chooses) and then dividing the ordinated swarm of data points with suitably placed partitions. The procedure is known as *ordination–space partitioning*. The term describes a large collection of methods since one can choose any one of a number of ways of doing the initial ordination, and then any one of a number of ways of placing the partitions. Gauch (1982) has reviewed the development of these methods. Collectively, they constitute a battery of exceedingly powerful procedures for interpreting ecological data. They yield an ordination and a classification simultaneously, and the classification, being divisive, avoids the disadvantage of agglomerative classifications previously described.

It should be noticed, however, that ordination–space partitioning is a much more "rough and ready" method of classifying quadrats than the agglomerative methods described in Chapter 2. Even so, it is probably adequate for most, if not all, ecological applications. And, as so often happens in efforts to interpret ecological data, one is faced with the perennial problem of choosing, judiciously, one of a large number of possible and only slightly different procedures.

In the following sections, we consider a few representative methods.

5.2. CONSTRUCTING AND PARTITIONING A MINIMUM SPANNING TREE

It was remarked previously that one (and only one) of the agglomerative methods of classification can be done in reverse. The method is nearest-neighbor clustering (see page 15), also known as single linkage clustering.

To do a nearest-neighbor classification divisively, the n data points are first plotted in s-space (n is the number of quadrats to be classified, and s the number of species); this is equivalent to doing an ordination of the data with no reduction in dimensionality.

The points of the swarm are then linked by a *minimum spanning tree*. A spanning tree is a set of line segments linking all the n points in the swarm in such a way that every pair of points is linked by one and only one path (i.e., a line segment, or sequence of connected line segments). None of the paths form closed loops. The length of the tree is the sum of the lengths of its constituent line segments. The minimum spanning tree of the swarm is the spanning tree of minimum length. (Note: Do not confuse a spanning tree with a tree diagram or dendrogram.)

Figure 5.1 shows a simple example with $n = 10$ and $s = 2$. The coordinates of the 10 data points in two-space are given in Data Matrix #1 (see Table 2.1, page 18), and the swarm of points is identical with the swarm in Figure 2.3*a* (page 17). For clarity, the data points are here labeled with the letters A, B, ..., J instead of with the numerals 1, 2, ..., 10. The length of each line segment is shown in the figure.

Partitioning the Tree

Nearest-neighbor classification by the divisive method is now done by cutting, in succession, first the longest link in the tree, then the second longest link, then the third longest, and so on. The process is illustrated, up to the penultimate step, by the sequence of dendrograms in Figure 5.2. The ultimate step, that of cutting the shortest link of all, yields a dendrogram identical with that in Figure 2.3*b*.

The procedure is very easy to carry out when, as in the example, there are only two species; for then the data swarm can be plotted, and the minimum spanning tree drawn, on a two-dimensional sheet of paper. Let us now

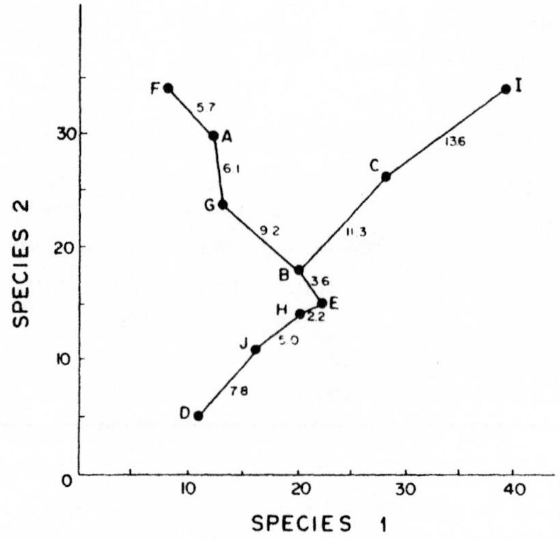

Figure 5.1. The points of Data Matrix #1 linked by their minimum spanning tree. The coordinates of the quadrat points (here labeled with the letters A, B, . . . , J) are given in Table 2.1. The distance between every pair of joined points is also shown.

describe the procedure in such a way that it is applicable whatever the value of s. It is still convenient to use Data Matrix #1 as an example.

If the data swarm is many-dimensional and hence unvisualizable, the line segments forming the minimum spanning tree must be found by inspecting the $n \times n$ distance matrix showing the distance between every pair of points.

The distance matrix for Data Matrix #1 is given in the upper panel of Table 5.1. It is identical with the distance matrix in Table 2.1 (page 18) except for two changes. In Table 5.1, the quadrats have been labeled with letters instead of numerals, as explained. And some of the distances in the matrix have been given superscript numbers; these label the segments of the minimum spanning tree in the order in which they are found. They are found as follows. (The method is due to Prim, and is described in Gower and Ross, 1969; Rohlf, 1973; Ross, 1969.)

The first segment of the tree corresponds to the shortest distance in the table. The shortest distance is $2.2 = d(\mathrm{E, H})$, the length of segment EH; therefore, superscript 1 is attached to this distance. We next find, by searching the rows and columns headed E and H, the shortest distance linking a third point to either E or H. It is the distance $d(\mathrm{E, B}) = 3.6$;

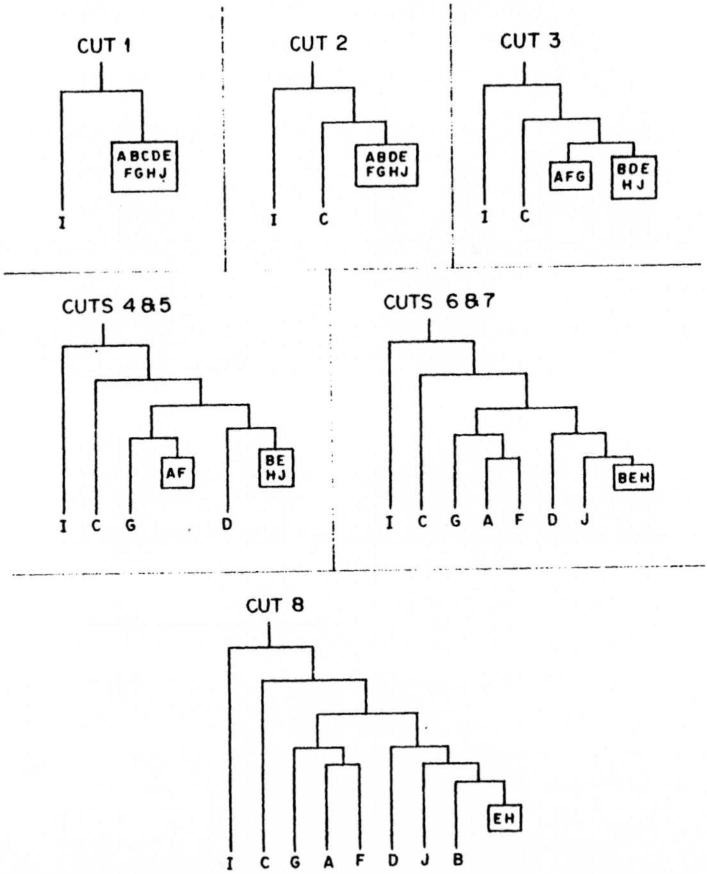

Figure 5.2. Stages in the construction of the dendrogram giving a nearest-neighbor classification of Data Matrix #1. Each successive cut permits a division to be made. The final (ninth) cut gives the complete dendrogram, which is shown in Figure 2.3*b*.

therefore, superscript 2 is attached to this distance. We next find, by searching the rows and columns headed E, H, and B, the shortest distance linking a fourth point to any of E, H, or B. It is the distance $d(\text{H}, \text{J}) = 5.0$; therefore, superscript 3 is attached to this distance. Notice that although $d(\text{B}, \text{H}) = 4.0$ is shorter than $d(\text{H}, \text{J})$, segment BH is not admissible as part of the minimum spanning tree as it would form a loop BHE and loops are not permitted.

Continuing in the same way, all the $n - 1 = 9$ segments needed to complete the tree are found. They are listed, with their lengths and in the

TABLE 5.1. A DIVISIVE NEAREST-NEIGHBOR CLASSIFICATION OF DATA MATRIX #1.[a]

The distance matrix:[b]

	A	B	C	D	E	F	G	H	I	J
A	0	14.4	16.5	25.0	18.0	5.7^{7}	6.1^{6}	17.9	27.3	19.4
B		0	11.3^{8}	15.8	3.6^{2}	20.0	9.2^{5}	4.0	24.8	8.1
C			0	27.0	12.5	21.5	15.1	14.4	13.6^{9}	19.2
D				0	14.9	29.2	19.1	12.7	40.3	7.8^{4}
E					0	23.6	12.7	2.2^{1}	25.5	7.2
F						0	11.2	23.3	31.0	24.4
G							0	12.2	27.9	13.3
H								0	27.6	5.0^{3}
I									0	32.5
J										0

The lengths of the segments of the minimum spanning tree, in the order in which they were found:

1: $d(\text{E},\text{H}) = 2.2$; 2: $d(\text{E},\text{B}) = 3.6$; 3: $d(\text{H},\text{J}) = 5.0$;
4: $d(\text{J},\text{D}) = 7.8$; 5: $d(\text{B},\text{G}) = 9.2$; 6: $d(\text{G},\text{A}) = 6.1$;
7: $d(\text{A},\text{F}) = 5.7$; 8: $d(\text{B},\text{C}) = 11.3$; 9: $d(\text{C},\text{I}) = 13.6$.

Diagram of the minimum spanning tree, constructed from the segments whose lengths are given above. The segments are labeled 1, 2, ..., 9 from longest to shortest, showing the order in which they are to be cut.

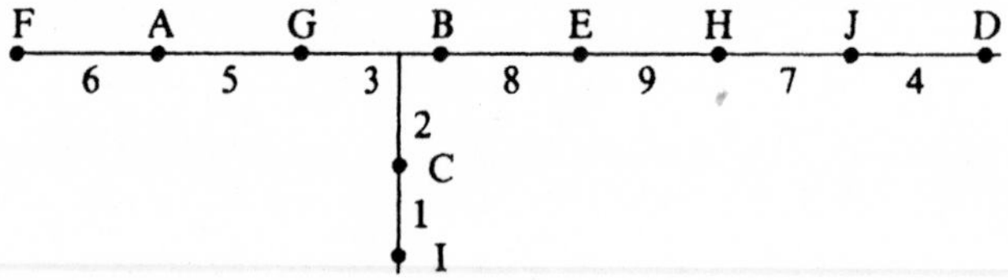

[a]See Table 2.1 and Figures 5.1 and 5.2.
[b]The row and column headings refer to the quadrats.

order in which they were found, in the center panel of Table 5.1. Observe that they were not discovered in order of increasing length; some of the later-found segments are shorter than some found earlier. Now that the segments have been found, it is easy to draw a two-dimensional diagrammatic representation of the minimum spanning tree, as has been done in the bottom panel of Table 5.1. The segments are linked together in the order in which they were identified; they have been assigned ranks according to their lengths with 1 for the longest up to 9 for the shortest.

To obtain a nearest-neighbor classification from the minimum spanning tree, it remains to cut the tree's segments, one after another, beginning with the largest. This partitioning process has already been demonstrated in Figure 5.2. Of course, with a many-dimensional data swarm which cannot be plotted and partitioned as was the two-dimensional swarm in Figure 5.1, the partitioning must be done on the diagrammatic minimum spanning tree constructed as shown in Table 5.1. The diagram can always be drawn in two dimensions, regardless of the value of s. In practice, of course, all the steps in the classification can be done by computer; a program has been given by Rohlf (1973). The foregoing description of the method explains its principles.

Clarifying an Ordination with a Minimum Spanning Tree

When s-dimensional data (with large s) have been ordinated in two-space, for example by PCA, there is obviously always a risk that two points that are far apart in s-space will appear close together in two-space. This is particularly likely to happen if the data have a nonlinear structure (see Chapter 4, Section 4.5). To avoid being misled by the spurious proximity of points that are, in fact, widely dissimilar, it often helps to draw the minimum spanning tree on an ordination diagram. Then the fact that apparently similar points are not linked by a segment of the tree makes it obvious that the similarity is only apparent.

Figure 5.3 shows an example. The data are from Delaney and Healy (1966) and the analysis from Gower and Ross (1969). The purpose of the research was to investigate the relationships among 10 isolated populations of shrews of the genus *Crocidura* on the basis of several skull measurements. Ordinating the data and reducing their dimensionality to 2 allows the 10 data points to be plotted in two-space, as shown in the figure. The populations belong to two groups of five each. One group comes from five of the Scilly Islands, off the southwestern tip of England; the other group comes from four of the Channel Islands off the north coast of France, and from Cap Gris Nez on the French mainland. If the ordination did not show the minimum spanning tree, it would be natural to conclude that two of the

Channel Island populations (those from Jersey and Sark, labeled J and S) were closely similar. In fact, as the minimum spanning tree shows, they are more similar to one of the Scilly Island populations than to each other.

Thus a minimum spanning tree is a useful adjunct to an ordination and can be helpful in preventing misinterpretations. When the number of points being ordinated is fairly low, then the minimum spanning tree can be shown as part of the ordination, as in Figure 5.3. When there are a large number of points, a diagram showing the tree as well as the points may be too confusing to be useful as a final portrayal of one's results; even so,

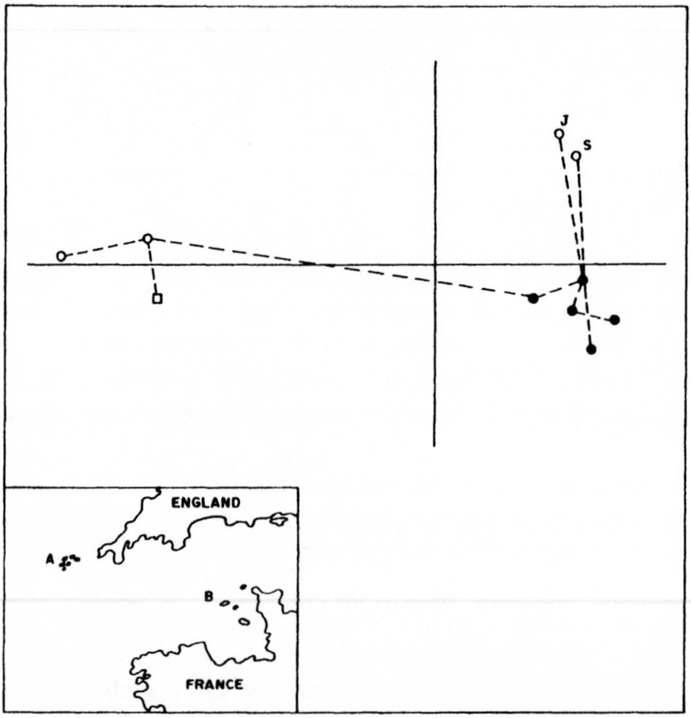

Figure 5.3. A two-dimensional ordination of a nine-dimensional data swarm showing the minimum spanning tree (dashed line). The original data were skull measurements on shrews, collected from 10 local populations. Five of the populations (solid dots) are from different islands in the Scilly Islands. The hollow symbols refer to four of the Channel Islands (circles) and Cap Gris Nez (square); J = Jersey; S = Sark. (Adapted from Gower and Ross, 1969.) The inset map shows the locations of the two island groups (A = Scilly Is., B = Channel Is.); Cap Gris Nez is on the north coast of France, far to the east.

examination of the tree should form part of the investigation so that spurious "proximities" will not go unnoticed.

5.3. PARTITIONING A PCA ORDINATION

It was remarked previously that the partitioning (according to definite rules) of a minimum spanning tree yields a nearest-neighbor classification. However, nearest-neighbor classifications have many defects (page 22). Thus although minimum spanning trees are useful as aids to the correct interpretation of ordination patterns, the partitioning of them is not (usually) a good way of doing a divisive classification. There are several better methods of partitioning a data swarm, and probably the most straightforward of all is Lefkovitch's (1976) method.

Lefkovitch's Partitioning Method

In geometric terms, the method consists in first ordinating the data in s-space (i.e., with no reduction in dimensionality) by means of a centered PCA; this ensures that the coordinate frame has its origin at the centroid of the data swarm, and its axes aligned with the principal axes of the swarm. Then the first division is made by "breaking" the first axis at the centroid, the second division by breaking the second axis at the centroid, and so on.

EXAMPLE. A simple, artificial example serves as a demonstration. We classify six quadrats that contain, together, eight species. The data are in Data Matrix #14, in the top panel of Table 5.2. The columns are headed with the letters A,...,F, which are the quadrat labels. An unstandardized, centered PCA was carried out on these data in order to determine the principal component scores (the coordinates measured on the principal axes) of the six data points. These coordinates, rounded to one decimal place, are given as the columns of matrix **Y** in the center panel of Table 5.2. Notice that after the PCA, the six points are ordinated in five-space. Hence **Y** has five rows; the eigenvalue corresponding to each row is shown on the right. The smallest (fifth) eigenvalue is much smaller than the others, and the coordinates on the fifth axis are all zero after rounding to one decimal place.

TABLE 5.2. CLASSIFICATION OF SIX QUADRATS BY LEFKOVITCH'S METHOD.

Data Matrix #14:

$$\mathbf{X} = \begin{array}{c} \begin{array}{cccccc} \mathrm{A} & \mathrm{B} & \mathrm{C} & \mathrm{D} & \mathrm{E} & \mathrm{F} \end{array} \\ \begin{pmatrix} 9 & 10 & 4 & 0 & 0 & 1 \\ 28 & 25 & 3 & 1 & 1 & 0 \\ 37 & 39 & 50 & 40 & 45 & 46 \\ 14 & 15 & 65 & 50 & 42 & 40 \\ 2 & 1 & 20 & 8 & 10 & 0 \\ 8 & 11 & 19 & 15 & 12 & 0 \\ 1 & 3 & 21 & 10 & 11 & 50 \\ 7 & 7 & 30 & 25 & 24 & 23 \end{pmatrix} \end{array}$$

The matrix of coordinates after PCA (the principal component scores):

$$\mathbf{Y} = \begin{array}{c} \begin{array}{cccccc} \mathrm{A} & \mathrm{B} & \mathrm{C} & \mathrm{D} & \mathrm{E} & \mathrm{F} \end{array} \\ \begin{pmatrix} -36.8 & -33.8 & 30.5 & 11.8 & 7.5 & 20.9 \\ 0.3 & 0.8 & -14.3 & -11.5 & -7.2 & 31.9 \\ 2.2 & 2.2 & 9.5 & -7.2 & -6.8 & 0.1 \\ -0.5 & 0.1 & 0.2 & -3.9 & 4.3 & -0.3 \\ 0.0 & 0.0 & 0.0 & 0.0 & 0.0 & 0.0 \end{pmatrix} \end{array} \quad \begin{array}{l} \\ \lambda_1 = 676.1 \\ \lambda_2 = 234.1 \\ \lambda_3 = 32.9 \\ \lambda_4 = 5.6 \\ \lambda_5 = 1.6 \end{array}$$

The matrix of signs:

$$\mathbf{Y}_s = \begin{array}{c} \begin{array}{cccccc} \mathrm{A} & \mathrm{B} & \mathrm{C} & \mathrm{D} & \mathrm{E} & \mathrm{F} \end{array} \\ \begin{pmatrix} - & - & + & + & + & + \\ + & + & - & - & - & + \\ + & + & + & - & - & + \\ - & + & + & - & + & - \\ \cdot & \cdot & \cdot & \cdot & \cdot & \cdot \end{pmatrix} \end{array}$$

Figure 5.4*a* shows the data swarm projected onto two-space; equivalently, it is a two-dimensional PCA ordination of the data. The coordinates of the points are given by the first two rows of **Y**.

Matrix $\mathbf{Y}_s$ in the bottom panel of the table gives the signs of the corresponding elements in **Y**, which is all the information required for carrying out the classification. To make the first division, consider the first row of $\mathbf{Y}_s$. We see that A and B both have minus signs, whereas C, D, E, and F all have plus signs. Hence the first division is into the two classes (A, B) and (C, D, E, F). The division is shown diagrammatically in the first dendrogram in Figure 5.4*b*. That this should be the first division is also obvious

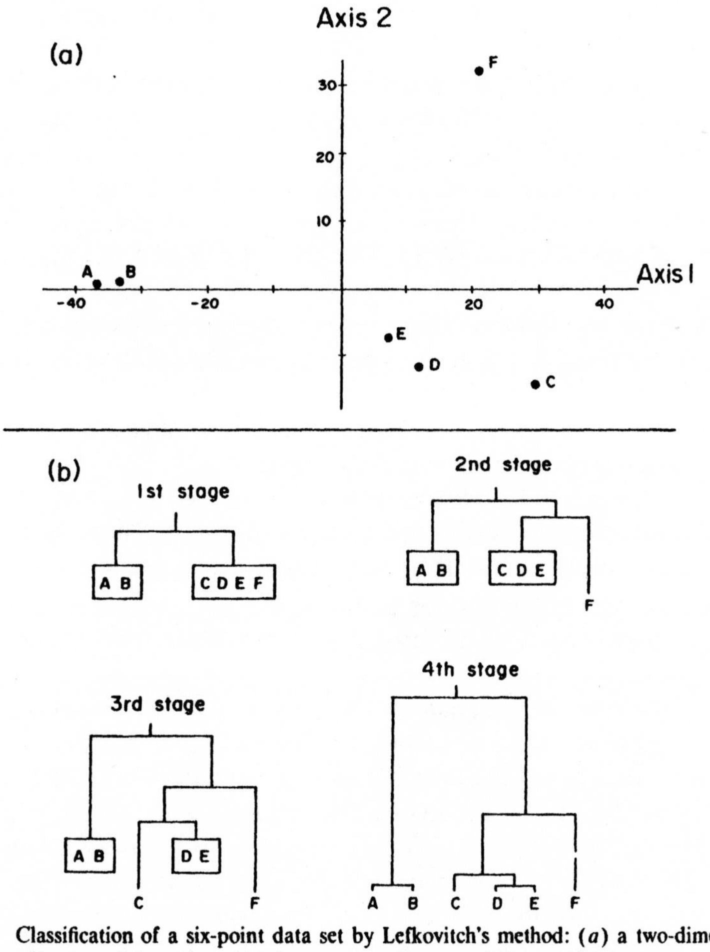

Figure 5.4. Classification of a six-point data set by Lefkovitch's method: (*a*) a two-dimensional PCA ordination of the points; (*b*) the sequence of stages of the partitioning.

from the scatter diagram in Figure 5.4*a* where it is seen that points A and B have negative coordinates, and the remaining points positive coordinates, on axis 1.

The second division, likewise, can be made either by examining row 2 of $\mathbf{Y}_s$, or from a glance at the scatter diagram. On axis 2, point F has a positive coordinate and points C, D, and E have negative coordinates. Hence the second division splits class (C, D, E, F). The three classes now recognized are, therefore, (A, B), (C, D, E), and (F) as shown in the second dendrogram in Figure 5.4*b*.

To do the third division we must consult $\mathbf{Y}_s$, since Figure 5.4*a* is only two-dimensional. (The geometric implications can still be visualized at this stage, however.) From row 3 of $\mathbf{Y}_s$ it is seen that, on the third axis A, B, C, and F have positive coordinates, and D and E have negative coordinates. We therefore separate those points that have different signs *for the first time*, namely, (C) from (D, E). This gives the third dendrogram in Figure 5.4*b*.

The two two-member classes present at the third division are both split at the fourth division: (A, B) splits into (A) and (B); (D, E) splits into (D) and (E). This is clear from row 4 of $\mathbf{Y}_s$, but now it cannot be visualized geometrically since we are concerned with the coordinates of the points on a fourth axis perpendicular to the other three.

Observe that points that have become separated at an early stage of the classification cannot become reunited at a later stage. For example, B and C become members of a different class at the first division because they are on opposite sides of the centroid as measured along axis 1. The fact that they are on the same side as measured along axes 3 and 4 (see rows 3 and 4 of $\mathbf{Y}_s$) is irrelevant.

In the final dendrogram, the heights of the first, second,..., nodes (counting from the top downwards) are proportional to the first, second,..., eigenvalues. Hence the height of a node shows the relative length of the axis that was "broken" at the division forming the node.

There is, of course, no need to continue the subdivision process right to the end, leaving every individual point (quadrat) isolated in a one-member cluster. One usually wishes to stop at a stage that leaves "real" clusters undivided; for instance, in the example in Figure 5.4, one might regard (A, B) and (D, E) as true, natural clusters and treat classification as complete at the third stage. This is one of the advantages of a divisive classification: the subdivision process need not be continued beyond the stage at which all "truly" separate clusters have been separated. The associated disadvantage is that a decision must be made as to how a "real" cluster shall be defined; equivalently, a rule has to be devised for deciding when the sequence of subdivisions should stop. Such a rule is unavoidably arbitrary and there are several possible criteria for deciding when subdivision has been carried far enough (Pielou, 1977). Discussion of these so-called *stopping rules* is beyond the scope of this book.

Lefkovitch's partitioning method has the merits that it is straightforward and can be counted upon to bring about the separation of all the natural clusters that may be present in a body of data. Its chief disadvantage is that it sometimes leads to the unwanted splitting of a cluster of closely spaced points that "should" be left together. This happens when such a cluster is transfixed by one of the principal axes. An example is given in Exercise 5.2. Therefore, the results of a classification by the method have to be tested to ensure that no such unwanted splits have occurred, and to undo them if they have. Lefkovitch (1976) suggests possible ways of doing the testing; the need for tests unfortunately detracts from the simplicity of the partitioning method.

Noy-Meir's Partitioning Method

This method (Noy-Meir, 1973b) predates Lefkovitch's but is described here in second place because it is not quite so simple. As with Lefkovitch's method, the data are first ordinated (with no reduction in dimensionality) by PCA; the principal axes are then broken in two, one after another, starting with the first. Noy-Meir's method differs in the way in which the "break point" is chosen for each break. It is so placed as to make the sum of the (within-group) variances of the principal component scores of the groups of points on either side of the break point as small as possible.

EXAMPLE. Let us classify Data Matrix #14 again, using Noy-Meir's method. The steps in the process are shown in full in Table 5.3.

Each axis is broken in turn. Each possible break point along each axis is tested in turn so that the correct point may be determined. For instance, consider the first axis, and the result of breaking it between points D and E.

The scores of the $n_1 = 4$ points (A, B, C, and D) to the left of this break point are

$$(y_1, y_2, y_3, y_4) = (-36.8, -33.8, 30.5, 11.8)$$

with variance

$$\frac{1}{n_1 - 1}\left\{\sum_{i=1}^{4} y_i^2 - \frac{1}{n_1}\left(\sum_{i=1}^{4} y_i\right)^2\right\} = 1121.98.$$

TABLE 5.3. CLASSIFICATION OF SIX QUADRATS BY NOY-MEIR'S METHOD.[a]

First Break

The scores on the first principal axis are

Point:	A	B	C	D	E	F
Score:	−36.8	−33.8	30.5	11.8	7.5	20.9

Break Between	Within-Group Variance		Sum of Variances
	Left Group	Right Group	
A and B	0	608.17	608.17
B and C	4.50	104.31	108.81*
C and D	1445.46	46.81	1492.27
D and E	1121.98	89.78	1211.76
E and F	883.97	0	883.97

The smallest sum is marked with an asterisk.
Make the break between B and C.

Second Break

The scores on the second principal axis are

Point:	A	B	C	D	E	F
Score:	0.3	0.8	−14.3	−11.5	−7.2	31.9

Break Between	Within-Group Variances		Sum of Variances
	Left Group	Right Group	
A and B	0	351.70	351.70
C and D	73.57	571.81	645.38
D and E	61.65	764.41	826.06
E and F	46.45	0	46.45*

Make the break between E and F.

Third Break

The scores on the third principal axis are

Point:	A	B	C	D	E	F
Score:	2.2	2.2	9.5	−7.2	−6.8	0.1

Break Between	Within-Group Variance		Sum of Variances
	Left Group	Right Group	
A and B	0	48.05	48.05
C and D	17.76	16.84	34.60*
D and E	46.85	23.81	70.66

Make the break between C and D.

[a]See Table 5.2 for the Data Matrix **X** and the matrix of principal component scores **Y**.

The scores of the $n_2 = 2$ points (E and F) to the right of the break point are

$$(y_5, y_6) = (7.5, 20.9)$$

with variance

$$\frac{1}{n_2 - 1}\left\{\sum_{i=5}^{6} y_i^2 - \frac{1}{n_2}\left(\sum_{i=5}^{6} y_i\right)^2\right\} = 89.78.$$

The sum of these two variances is 1211.76. (It should now be clear how all the entries in the table are computed.)

It is seen that the smallest sum of variances is obtained when the break on this axis is made between points B and C. Hence the first division of the classification is into the classes (A, B) and (C, D, E, F).

The second division is made by breaking the second axis. In this case the break comes between the points E and F. Therefore, the three classes recognized after the second division are (A, B), (C, D, E), and (F). Likewise, the four classes after the third division are (A, B), (C), (D, E), and (F). The ultimate step, which needs no computation, is to break A from B and D from E.

It should be noticed that the sequence of breaks is identical with that yielded by Lefkovitch's method.

In the example, each axis was broken exactly once. In another version of the method, an axis may be broken more than once if such a break gives a smaller sum of variances than would the breaking of a hitherto unbroken axis. Details are given by Noy-Meir (1973b). He also discusses applications of the method to data ordinated by other forms of PCA (besides unstandardized centered PCA as used here). And he suggests possible stopping rules. The method does not lead to unwanted splitting of tight clusters of points as Lefkovitch's method sometimes does when a cluster happens to be skewered by one of the principal axes (see Exercise 5.2).

It is interesting to note the resemblance of the method to minimum variance clustering (page 32). It is not true to say, however, that Noy-Meir's partitioning method amounts to minimum variance clustering done "in reverse." Thus one does not examine every possible division of the points into two groups to find which gives the minimum sum of within-group variances; such a procedure would be impracticable because of the excessive amount of computation required (page 204). The only divisions examined are those corresponding to breaks of the principal component axes; hence, given n points, there are only $n - 1$ possible break points on each axis.

5.4. PARTITIONING RA AND DCA ORDINATIONS

Any partitioning method devised for application to PCA or PCO ordinations can, of course, be applied to RA and DCA ordinations as well, and vice versa.

Hill (1979b; and see Hill, Bunce, and Shaw, 1975) developed a partitioning procedure that was applied to RA ordinations, but there is no reason why it should not be used with PCA and PCO ordinations. In principle, it consists in carrying out a one-dimensional RA ordination and breaking the axis at the centroid so as to divide the data points into two classes. Each of the two classes is then itself split, in exactly the same way, to give a total of four classes; then each of the four classes is split to give eight classes, and so on.

The method is known as two-way indicator species analysis; a computer program for doing it, called TWINSPAN, is available (Hill, 1979b) and, as its author comments, it is "long and rather complicated." This is because, at each step, the required one-dimensional RA ordination is first done in the ordinary way to give a "crude" partitioning of the data points; it is then redone (at least once and sometimes twice) with the species quantities weighted in such a way as to emphasize the influence of especially useful diagnostic species (i.e., of differential, or "indicator," species) identified by the first ordination.

These (and other) refinements are thought to make the classification more natural by ensuring that "indifferent" species (those that are not diagnostic of true natural classes) do not affect the results. However, the price of such refinements is the loss of simplicity. Whenever a simple, basic method of analysis is refined and elaborated, the number of possible modified forms of the original method increases exponentially and choosing among them becomes increasingly subjective.

It is worth considering an example (in Figure 5.5) of a TWINSPAN analysis in order to demonstrate how clearly the results can be displayed. One can have the best of two worlds (classification and ordination) by presenting a two or three-dimensional ordination of the data under investigation, and then drawing the partitions that yield the classification directly on the ordination scatter diagram. To complete the representation, the classification dendrogram is given as well.

Figure 5.5 shows the result of an ordination-plus-classification of vegetation. It is adapted from a figure in Marks and Harcombe (1981). The scatter diagram shows a two-dimensional RA ordination of 54 sample plots representing the range of natural vegetation in the coastal plain of southeastern Texas. The data matrix was also classified, using the TWINSPAN program, and gave the classification dendrogram shown as an inset on the graph. The four groups of sample plots separated in the classification were then outlined and labeled on the ordination.

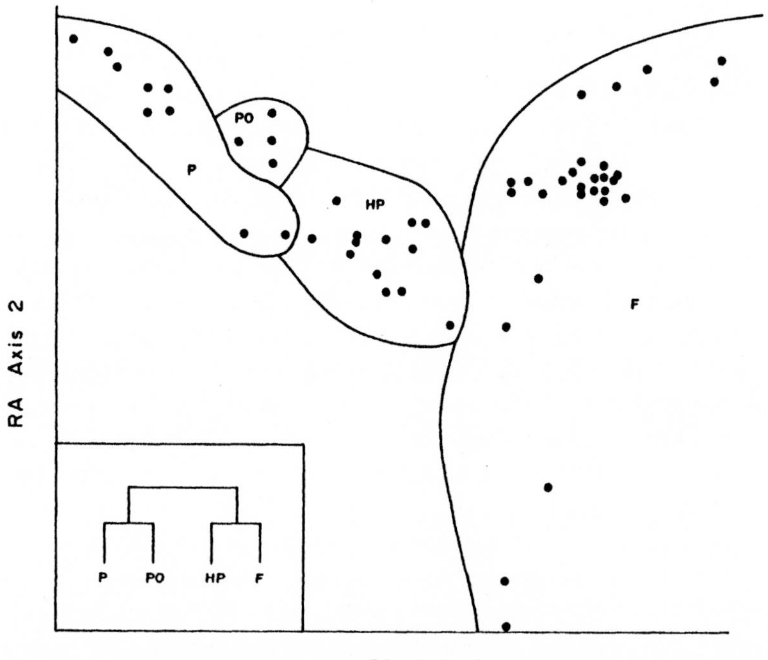

Figure 5.5. A two-dimensional ordination and a TWINSPAN classification of data on the vegetation of 54 sample plots of vegetation in southeastern Texas. The vegetation classes distinguished are: P, sandhill and upland pine forest and wetland pine savanna; PO, pine-oak forest on upper slopes; HP, hardwood and pine forest; F, flatland and floodplain hardwood forest, and wetlands and shrub thickets. [Adapted from Marks and Harcombe (1981). In the original paper the classification is carried further and the ordination diagram is partitioned into 10 classes rather than merely 4.]

It should be noticed that when an ordination and a classification are done simultaneously, it becomes possible to represent the classification dendrogram in the most natural way possible. Thus, consider Figure 5.5. If the only analysis to which the data had been subjected had been a classification into four classes, the resultant dendrogram, thought of as a mobile capable of swiveling at every node, could have been drawn in any one of eight ways; for instance, one of the other possible versions (in addition to the one shown in Figure 5.5) is the following.

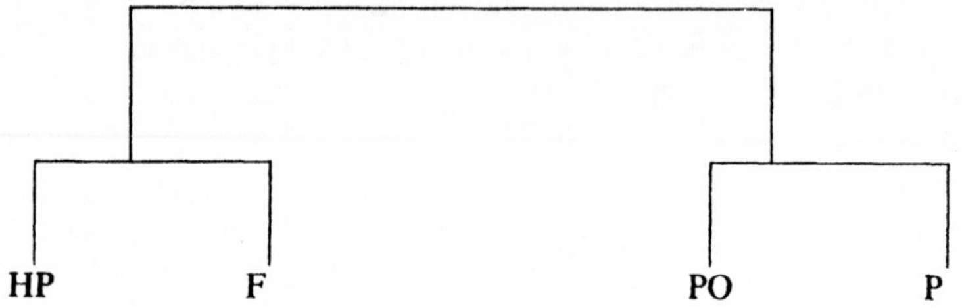

However, only two of the ways (that in Figure 5.5 and its mirror image) show that, for example, HP is closer (more similar) than F to PO, and that the greatest separation (dissimilarity) is between F and P. It should now be clear that the numerous possible ways of drawing a dendrogram are not all equally informative. One of the merits of the TWINSPAN program is that it arranges the dendrogram's branches in a way that puts similar points close to each other, so far as is possible in a two-dimensional representation.

Since a TWINSPAN classification entails a new one-dimensional RA ordination at each step, the same result would be obtained if DCA ordinations were used. This is because the order of the points on the first axis is the same with a DCA as with an RA ordination. Partitioning a two-dimensional DCA ordination is yet another way of doing a divisive classification; it has been proposed and demonstrated by Gauch and Whittaker (1981) who gave the procedure the name DCASP. The partitioning is done subjectively. Partitions are drawn through parts of the scatter diagram where the data points can be seen to be sparse and therefore the method is unlikely to make "false" divisions. But there is a risk that some "true" divisions may escape notice. Data points may appear close in a two-dimensional diagram even though they are far apart in many-dimensional space (for an example, see Figure 5.3). It may be feasible to guard against being misled by drawing the minimum spanning tree of the data swarm that is to be partitioned.

EXERCISES

5.1. The following distance matrix gives the pairwise distances between points in a swarm of 10 points in nine-space. The points are labeled A, B, ..., J. Find the segments of the minimum spanning tree and list them with their lengths in the order in which they were found (as in the center panel of Table 5.1). Draw a diagram of the minimum spanning tree.

	A	B	C	D	E	F	G	H	I	J
A	0	1.88	2.33	2.26	1.74	2.93	3.30	10.73	8.83	8.57
B		0	2.54	2.97	2.05	4.00	4.52	10.89	9.09	8.78
C			0	3.22	1.54	4.01	4.10	11.28	9.66	9.21
D				0	2.68	4.51	3.46	10.01	8.20	8.24
E					0	3.84	3.56	10.54	9.04	8.64
F						0	3.37	10.99	9.00	8.74
G							0	10.44	8.96	9.07
H								0	3.27	3.77
I									0	3.00
J										0

5.2. Refer to Table 5.2. If Data Matrix #14 is altered by putting $x_{61} = x_{62} = 0$, it is found that the matrix of principal component scores becomes

$$\mathbf{Y} = \begin{pmatrix} -37.5 & -34.8 & 32.2 & 13.2 & 8.4 & 18.4 \\ -1.2 & 0.6 & -12.5 & -11.7 & -7.7 & 32.5 \\ 2.5 & 1.8 & 9.5 & -7.0 & -6.4 & -0.3 \\ -0.7 & 0.4 & 0.2 & -3.8 & 4.3 & -0.3 \\ -1.3 & 1.4 & 0.0 & 0.2 & -0.2 & -0.1 \\ -0.0 & 0.0 & 0.0 & 0.0 & -0.0 & 0.0 \end{pmatrix}$$

Do a divisive classification of these data by Lefkovitch's method. Plot the two-dimensional ordination.

5.3. Carry out a divisive classification of the data described in Exercise 5.2 using Noy-Meir's method. Stop when four classes have been distinguished.

Chapter Six

Discriminant Ordination

6.1. INTRODUCTION

The data matrices that have been described, ordinated, and classified so far in this book have all been treated in isolation. It has been assumed that an investigator has only one data matrix to interpret at any one time. We now suppose that several data matrices are to be interpreted jointly. It is desired to ordinate all of them together, that is, in a common coordinate frame, and an ordination method is wanted that emphasizes as much as possible the contrasts among them.

Here are several examples of the kinds of investigations in which joint ordinations are helpful.

1. Suppose one were investigating the emergent vegetation (or the benthic invertebrate fauna, or the diatom flora) of several lakes. The data would consist of several data matrices, one from each lake.
2. One might be sampling the insect fauna (or some taxonomic subset of it) in wheat fields in July in several successive years. Then the data would consist of several data matrices, one for each year.
3. One might be comparing environmental conditions in several geographically separate regions. Within each region, a number of environmental variables are measured in each of a number of "quadrats" (or other sampling stations) and the result is a data matrix summarizing conditions in

that region. Then the total data consist of several data matrices, one for each region.

It should now be clear that situations frequently arise in which it is desirable to ordinate several data matrices jointly. The researcher usually wants to know whether the separate data matrices (from different lakes, years, regions, or whatever it may be) differ from one another, and may do a multivariate analysis of variance to judge, objectively, whether they do. But independently of any statistical tests that may be done, it is clearly advantageous to be able to *see*, in a two-dimensional ordination on a sheet of paper, how the several sets of data are interrelated.

A way of achieving this is to ordinate all the data matrices jointly by means of a *discriminant ordination* (Pielou, unpublished). Before the method is described, we devote a section to necessary mathematical preliminaries.

6.2. UNSYMMETRIC SQUARE MATRICES

All ordination methods so far discussed in this book have entailed the eigenanalysis of a *symmetric* square matrix. A discriminant ordination requires that an *unsymmetric* square matrix be eigenanalysed. This section, therefore, describes some of the properties of unsymmetric square matrices and shows how they differ from symmetric matrices. In all that follows, the symbols **A** and **B** are used for symmetric and unsymmetric square matrices, respectively.

Factorization of an Unsymmetric Square Matrix

Recall that given a symmetric matrix **A**, one can always find an orthogonal matrix **U** and a diagonal matrix $\mathbf{\Lambda}$ such that

$$\mathbf{A} = \mathbf{U}'\mathbf{\Lambda}\mathbf{U}. \tag{6.1}$$

As always $\mathbf{U}'$ denotes the transpose of **U**.

To make the discussion clearer, we now change the symbols by putting $\mathbf{U}' = \mathbf{V}$; consequently, $\mathbf{U} = \mathbf{V}'$. Equation (6.1) now becomes

$$\mathbf{A} = \mathbf{V}\mathbf{\Lambda}\mathbf{V}'. \tag{6.2}$$

Let us rearrange (6.2). Postmultiplying both sides by **V** and using the fact that since **V** is orthogonal, $\mathbf{V}'\mathbf{V} = \mathbf{I}$, it is seen that

$$\mathbf{AV} = \mathbf{V\Lambda V}'\mathbf{V} = \mathbf{V\Lambda I}$$

or, more simply,

$$\mathbf{AV} = \mathbf{V\Lambda}. \tag{6.3}$$

The columns of **V** (which are the rows of **U**) are the eigenvectors of **A**, and the elements on the main diagonal of $\mathbf{\Lambda}$ are the eigenvalues of **A**. Indeed, (6.3) is the equation that defines the eigenvalues and eigenvectors of **A**.

An exactly analogous equation, namely,

$$\mathbf{BW} = \mathbf{W\Lambda} \tag{6.4}$$

defines the eigenvalues and eigenvectors of the unsymmetric matrix **B**. As before, the elements on the diagonal of the diagonal matrix $\mathbf{\Lambda}$ are the eigenvalues of **B** and the columns of **W** are its eigenvectors. But in this case

$$\mathbf{B} \neq \mathbf{W\Lambda W}'.$$

This is because **W** is not an orthogonal matrix; in symbols, $\mathbf{WW}' \neq \mathbf{I}$.

The Inverse of a Square Matrix

We now ask whether, given the matrix **W**, another matrix, to be denoted by $\mathbf{W}^{-1}$, can be found such that $\mathbf{WW}^{-1} = \mathbf{I}$. The answer is yes; $\mathbf{W}^{-1}$ is known as the *inverse* of **W**.

Indeed, apart from exceptions noted in Exercise 6.1, every square matrix has an inverse. Excluding the exceptions from consideration here, we can say that for any square matrix, say **M**, which may be symmetric or unsymmetric, another matrix, say $\mathbf{M}^{-1}$, can be found such that

$$\mathbf{MM}^{-1} = \mathbf{M}^{-1}\mathbf{M} = \mathbf{I}.$$

(Note that when a matrix is multiplied by its inverse, the order of the factors does not matter.) Moreover if, and only if, **M** is orthogonal, $\mathbf{M}^{-1} = \mathbf{M}'$. If **M** is not orthogonal, $\mathbf{M}^{-1} \neq \mathbf{M}'$.

Hence in deriving an equation of the form of (6.2) from (6.4) we postmultiply both sides of (6.4) by $\mathbf{W}^{-1}$ (not $\mathbf{W}'$) to get

$$\mathbf{BWW}^{-1} = \mathbf{W\Lambda W}^{-1}$$

or, more simply,

$$\mathbf{B} = \mathbf{W\Lambda W}^{-1} \tag{6.5}$$

The difference between (6.2) and (6.5) should be noted. It arises from the fact that **V**, whose columns are the eigenvectors of a symmetric matrix, is orthogonal. In contrast **W**, whose columns are the eigenvectors of an unsymmetric matrix, is nonorthogonal.

Finding the inverse of an orthogonal matrix presents no problem; one has only to write down its transpose. But finding the inverse of a nonorthogonal matrix requires very laborious computations. We do not describe the steps here. Clear expositions can be found in many books, for example, Searle (1966) and Tatsuoka (1971). For our purposes it suffices to note that usually (but see Exercise 6.1) an inverse for a square matrix can be found and that most computers have a function for obtaining it. As we see in the following, finding the inverse of a nonorthogonal matrix is one of the steps in carrying out a discriminant ordination.

As an illustrative example of a matrix and its inverse, suppose

$$\mathbf{M} = \begin{pmatrix} 2 & 4 & 3 \\ -3 & 2 & 1 \\ -1 & 3 & 2 \end{pmatrix}.$$

Its inverse is

$$\mathbf{M}^{-1} = \begin{pmatrix} 1 & 1 & -2 \\ 5 & 7 & -11 \\ -7 & -10 & 16 \end{pmatrix}.$$

The reader should check that $\mathbf{MM}^{-1} = \mathbf{M}^{-1}\mathbf{M} = \mathbf{I}$.

The Geometry of Orthogonal and Nonorthogonal Transformations

Let us now investigate the results of using a nonorthogonal matrix to *transform* another matrix.

First, recall what happens when an orthogonal matrix is used to effect a transformation. It was shown in Chapter 3 (page 94) that the effect of premultiplying a data matrix by an orthogonal matrix is, in geometric terms, to rotate the whole "data swarm" (the points defined by the transformed matrix) rigidly about the origin of the coordinate frame; equivalently, one can think of the data swarm as fixed and the transformation as rotating the coordinate frame (see Figure 3.3, page 97).

Now let us transform a data matrix by premultiplying it by a nonorthogonal matrix. As an example, let the matrix that is to be transformed be

$$\mathbf{X} = \begin{pmatrix} -1 & 1 & 1 & -1 \\ 1 & 1 & -1 & -1 \end{pmatrix},$$

whose columns give the coordinates of the corners of a square with center at the origin (see Figure 6.1*a*).

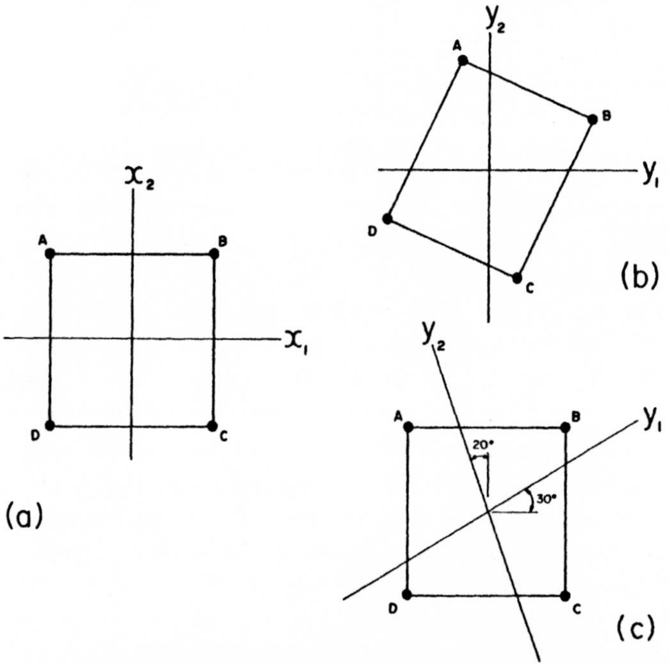

Figure 6.1. The effect of transforming data with a nonorthogonal matrix. (*a*) The original "data swarm," a square X; (*b*) and (*c*) show the transformation of X into TX in two different ways: (*b*) shows the axes unaltered but the points moved; (*c*) shows the points unaltered but the axes rotated, independently, through the angles shown. (Compare Figure 3.3.)

We now construct a 2×2 matrix **T** that is to be used to transform **X**. Let it be of the same form as the right-hand matrix in Equation (3.7) (page 101). That is, let us put

$$\mathbf{T} = \begin{pmatrix} \cos\theta_{11} & \cos\theta_{12} \\ \cos\theta_{21} & \cos\theta_{22} \end{pmatrix} \tag{6.6}$$

with $\theta_{12} = 90° - \theta_{11}$ and $\theta_{21} = 90° + \theta_{22}$. (Hence the squares of the elements in each row of **T** sum to unity.) But, unlike **U** in Equation (3.7), we choose different values for θ_{11} and θ_{22}. This ensures that $\mathbf{TT}' \neq \mathbf{I}$ and therefore **T** is nonorthogonal as required.

As a particular example, let

$$\mathbf{T} = \begin{pmatrix} \cos 30° & \cos 60° \\ \cos 110° & \cos 20° \end{pmatrix} = \begin{pmatrix} 0.866 & 0.500 \\ -0.342 & 0.940 \end{pmatrix}.$$

Then

$$\mathbf{TX} = \begin{pmatrix} -0.37 & 1.37 & 0.37 & -1.37 \\ 1.28 & 0.60 & -1.28 & -0.60 \end{pmatrix}.$$

Matrix **TX** is plotted in Figure 6.1 in two different ways. Figure 6.1*b* shows the points plotted in an "ordinary" coordinate frame with the axes perpendicular to each other. As may be seen, the square in Figure 6.1*a* has been distorted into the shape of a rhomboid, as well as being rotated; compare it with Figure 3.3*b*, in which the square, though rotated, is undistorted. In Figure 6.1*c* (which is comparable to Figure 3.3*c*), the square is the same shape and has the same orientation as in Figure 6.1*a*, but the coordinate axes have been rotated. In the present case (in contrast to Figure 3.3*c*) the axes have not been rotated as a rigid frame; instead, each axis has been rotated separately and the axes are no longer perpendicular. The angles between the new (y) axes and the old (x) axes are shown in the figure. Exercise 6.2 invites the reader to look even more closely at the geometry of nonorthogonal transformations.

Of course, not all nonorthogonal matrices are of the form of **T** in which the elements of each row are the direction cosines of a newly oriented coordinate axis with the axes not mutually perpendicular. Any matrix whose transpose is not identical with its inverse is nonorthogonal by definition. The effects of such matrices when used to transform other matrices have already been illustrated in Figure 3.2*b–e* (page 93). They rotate the axes (separately) and alter the scales as well. A matrix like **T** in (6.6) only rotates the axes; it leaves the scale of each axis unchanged.

Lastly, it should be noticed that although, for convenience, the preceding discussion deals with a data swarm in two-dimensional space and a 2×2 transformation matrix **T**, all the arguments can be extrapolated to as many dimensions as we wish.

Eigenanalysis of an Unsymmetric Square Matrix

The last matter to consider in this section on mathematical preliminaries is the eigenanalyses of unsymmetric square matrices.

The eigenvalues and eigenvectors of such a matrix, say **B**, are found by solving the set of equations implicit in Equation (6.4), namely,

$$\mathbf{BW} = \mathbf{W\Lambda}.$$

If **B** is an $n \times n$ matrix, then there are n eigenvalues (the elements on the main diagonal of Λ) and n eigenvectors (the columns of **W**). There are various ways of solving (6.4) but they are not described in this book. The principles are fully explained and illustrated in, for example, Searle (1966) and Tatsuoka (1971); applying them, except in artificially simple cases, entails heavy computations. Here we merely give a simple example to illustrate the results of such an eigenanalysis.

Suppose

$$\mathbf{B} = \begin{pmatrix} 67 & 92 & -148 \\ 18 & 29 & -42 \\ 41 & 59 & -92 \end{pmatrix}.$$

Then the equation $\mathbf{BW} = \mathbf{W\Lambda}$ is satisfied (as the reader should confirm) by putting

$$\Lambda = \begin{pmatrix} 3 & 0 & 0 \\ 0 & 2 & 0 \\ 0 & 0 & -1 \end{pmatrix} \quad \text{and} \quad \mathbf{W} = \begin{pmatrix} 2 & 4 & 3 \\ -3 & 2 & 1 \\ -1 & 3 & 2 \end{pmatrix}.$$

It follows that the elements on the main diagonal of Λ are the eigenvalues of **B**. The columns of **W** are proportional to the eigenvectors of **B**;

notice that the equation is still satisfied if the elements in any column of **W** are replaced by a constant multiple of the values shown. To normalize the eigenvectors or, which comes to the same thing, to put their elements in the form of direction cosines, it is necessary to divide through each column of **W** by the square root of the sum of squares of its elements. In the example, the normalized form of **W** is, therefore,

$$\begin{pmatrix} 0.5345 & 0.7428 & 0.8018 \\ -0.8018 & 0.3714 & 0.2673 \\ -0.2673 & 0.5571 & 0.5345 \end{pmatrix}.$$

6.3. DISCRIMINANT ORDINATION OF SEVERAL SETS OF DATA

Now that the groundwork has been laid, we consider how several sets of data may be simultaneously ordinated in a common coordinate frame in such a way as to separate the different swarms of points as widely as possible. The method is described here in recipe form because the underlying theory is beyond the scope of this book. Theoretical accounts may be found in, for example, Tatsuoka (1971) and Pielou (1977).

To illustrate the method, it is applied to real data. The data consist of values of 4 climatic variables observed at 14 weather stations in 3 geographic regions. The purpose of the analysis is to ordinate the stations on the basis of their climates.

In detail, the data are as follows. The locations of the weather stations are shown on the map in Figure 6.2, and the place names appear as column headings in Table 6.1. The three regions are: 1, the southern part of Yukon Territory in the Canadian boreal forest; 2, northern Alberta, also in the boreal forest but at a lower latitude; 3, southern Alberta in the Canadian prairies. The climatic variables are listed in a footnote to the table.

The data, which could be written out as three separate matrices, one for each region, have thus been brought together as the single large matrix, hereafter called **X**, shown in Table 6.1. The dashed vertical lines separate the three regions. The climatic observations for each station are shown in rows 3 through 6 of the matrix. It remains to explain the elements in rows 1 and 2.

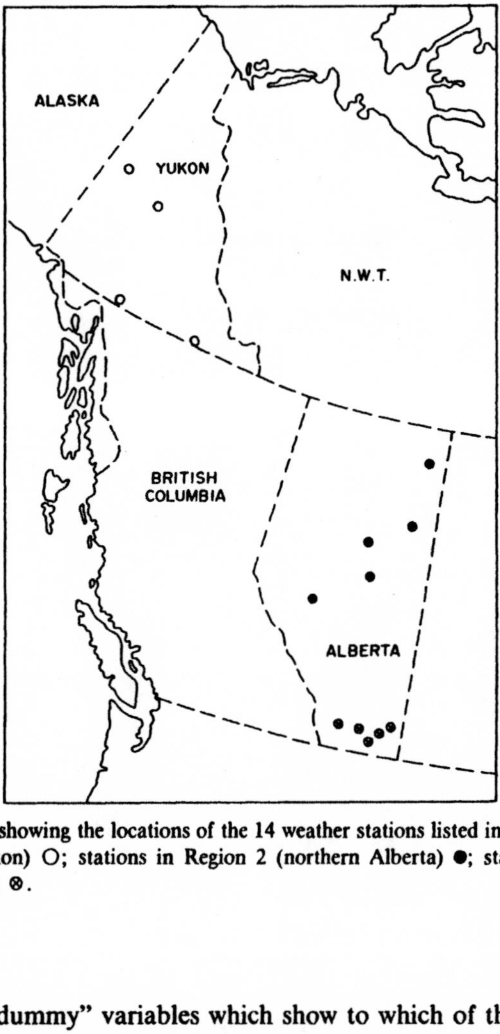

Figure 6.2. Map showing the locations of the 14 weather stations listed in Table 6.1. Stations in Region 1 (Yukon) O; stations in Region 2 (northern Alberta) ●; stations in Region 3 (southern Alberta) ⊗.

These are "dummy" variables which show to which of the three regions each of the stations belongs. As may be seen, every station has two dummy variables associated with it, x_1 and x_2. They are assigned as follows.

$$(x_1, x_2) = \begin{cases} (1,0) & \text{for all stations in Region 1} \\ (0,1) & \text{for all stations in Region 2} \\ (0,0) & \text{for all stations in Region 3.} \end{cases}$$

TABLE 6.1. DATA MATRIX GIVING THE VALUES OF 4 CLIMATIC VARIABLES AT 14 STATIONS IN 3 REGIONS.[a,b]

	Region 1 Yukon				Region 2 Northern Alberta					Region 3 Southern Alberta				
	Carcross	Dawson	Mayo	Watson Lake	Athabasca	Edson	Fort Chipewyan	Fort MacMurray	Wabasca	Foremost	Lethbridge	Fort McLeod	Medicine Hat	Raymond
x_1	1	1	1	1	0	0	0	0	0	0	0	0	0	0
x_2	0	0	0	0	1	1	1	1	1	0	0	0	0	0
x_3	−18.9	−29.4	−25.0	−20.0	−17.2	−12.8	−25.0	−22.8	−18.9	−11.1	−8.9	−8.3	−11.1	−8.3
x_4	12.8	15.6	14.4	14.4	15.6	15.6	16.7	16.1	15.6	20.0	17.8	18.3	20.6	18.9
x_5	11.5	13.8	10.1	17.2	14.7	13.1	11.5	14.8	10.3	1.1	11.9	11.5	9.8	12.9
x_6	11.3	18.3	18.4	23.0	31.9	34.2	20.4	30.1	31.8	29.1	26.2	26.6	22.8	26.2

[a] Data from "Climatic Summaries for Selected Meteorological Stations in the Dominion of Canada, Volume I," Meteorological Division, Department of Transport, Canada, Toronto, 1948.

[b] x_1 and x_2 are dummy variables; see text; x_3 and x_4 are daily mean temperatures in degrees C in January and July, respectively; x_5 and x_6 are precipitation in cm for October to March and April to September, respectively.

In general, if there were k regions, $k - 1$ dummy variables would be required to label all the stations. They would be

$$(x_1, x_2, \ldots, x_{k-1}) = \begin{cases} (1,0,\ldots,0) & \text{for all stations in Region 1} \\ (0,1,\ldots,0) & \text{for all stations in Region 2} \\ \cdots\cdots\cdots & \cdots\cdots\cdots\cdots\cdots\cdots \\ (0,0,\ldots,1) & \text{for all stations in Region } k-1 \\ (0,0,\ldots,0) & \text{for all stations in Region } k, \end{cases}$$

with $k - 1$ elements in each vector.

Thus, when there are n stations altogether grouped into k regions, and s variables are observed at each station, **X** has $s + k - 1$ rows and n columns. In the present case with $s = 4$, $k = 3$, and $n = 14$, **X** is a (6×14) matrix.

The operations to be carried out on matrix **X** are now described in numbered paragraphs.

1. Center and standardize the data as described earlier. That is, replace the (i, j)th element of **X**, x_{ij}, by $(x_{ij} - \bar{x}_i)/\sigma_i$ where $\bar{x}_i$ and σ_i are the mean and standard deviation of all the elements in the ith row. (Note: it makes no difference to the result whether the dummy variables are standardized. In the computations shown in Table 6.2 they are standardized. Every row must be centered.)
2. Postmultiply the matrix by its transpose to obtain the SSCP matrix **S**. **S** is shown in Table 6.2.
3. *Partition* **S** into four *submatrices* $\mathbf{S}_{11}$, $\mathbf{S}_{12}$, $\mathbf{S}_{21}$, and $\mathbf{S}_{22}$ as shown by the dashed lines. The four parts into which **S** has been divided are as follows: $\mathbf{S}_{11}$ is a $(k-1) \times (k-1) = 2 \times 2$ matrix giving the sums of squares and cross-products of the two dummy variables x_1 and x_2; $\mathbf{S}_{22}$ is the $s \times s = 4 \times 4$ matrix giving the sums of squares and cross-products of the observed variables x_3, x_4, x_5, and x_6; $\mathbf{S}_{12}$ is a $(k-1) \times s = 2 \times 4$ matrix whose elements are all sums of cross-products formed by multiplying one of the dummy variables by one of the observed variables; $\mathbf{S}_{21} = \mathbf{S}'_{12}$. (See Exercise 6.3.)
4. Obtain the inverses of $\mathbf{S}_{11}$ and $\mathbf{S}_{22}$, namely, $\mathbf{S}_{11}^{-1}$ and $\mathbf{S}_{22}^{-1}$. They are written out in full in Table 6.2.

TABLE 6.2. STEPS IN THE DISCRIMINANT ORDINATION OF THE DATA IN TABLE 6.1.

The SSCP matrix **S** is

$$\left(\begin{array}{rr|rrrr} 14.00 & -6.60 & -8.34 & -9.41 & 3.54 & -10.44 \\ -6.60 & 14.00 & -3.66 & -3.28 & 3.38 & 7.88 \\ \hline -8.34 & -3.66 & 14.00 & 9.05 & -4.22 & 5.89 \\ -9.41 & -3.28 & 9.05 & 14.00 & -7.20 & 4.78 \\ 3.54 & 3.38 & -4.22 & -7.20 & 14.00 & -0.41 \\ -10.44 & 7.88 & 5.89 & 4.78 & -0.41 & 14.00 \end{array}\right) = \left(\begin{array}{c|c} \mathbf{S}_{11} & \mathbf{S}_{12} \\ \hline \mathbf{S}_{21} & \mathbf{S}_{22} \end{array}\right).$$

The inverses of $\mathbf{S}_{11}$ and $\mathbf{S}_{22}$ are

$$\mathbf{S}_{11}^{-1} = \begin{pmatrix} 0.09184 & 0.04329 \\ 0.04329 & 0.09184 \end{pmatrix}; \qquad \mathbf{S}_{22}^{-1} = \begin{pmatrix} 0.13303 & -0.07552 & 0.00038 & -0.03014 \\ -0.07552 & 0.15650 & 0.05717 & -0.02004 \\ 0.00038 & 0.05717 & 0.10047 & -0.01677 \\ -0.03014 & -0.02004 & -0.01677 & 0.09046 \end{pmatrix}.$$

The product matrix **D** is

$$\mathbf{D} = \mathbf{S}_{22}^{-1}\mathbf{S}_{21}\mathbf{S}_{11}^{-1}\mathbf{S}_{12} = \begin{pmatrix} 0.40734 & 0.41975 & -0.25893 & -0.07804 \\ 0.53979 & 0.57768 & -0.29694 & 0.21171 \\ -0.00263 & -0.00195 & 0.00330 & 0.01165 \\ -0.05904 & -0.02255 & 0.11999 & 0.57397 \end{pmatrix}.$$

The eigenvalues of **D** are

$\lambda_1 = 0.96396$ and $\lambda_2 = 0.59832$.

The first two eigenvectors of **D** (normalized) are the rows of

$$\mathbf{W}'_{(2)} = \begin{pmatrix} -0.61003 & -0.78007 & 0.00494 & 0.13902 \\ -0.35471 & 0.02458 & 0.01981 & 0.93444 \end{pmatrix}.$$

5. Find the matrix product **D** defined as

$$\mathbf{D} = \mathbf{S}_{22}^{-1}\mathbf{S}_{21}\mathbf{S}_{11}^{-1}\mathbf{S}_{12}.$$

D is shown in Table 6.2.

6. Do an eigenanalysis of the unsymmetric square matrix **D**. The results are shown in Table 6.2. The number of nonzero eigenvalues is always the lesser of s (the number of variables of interest) and $k - 1$ (where k is the number of groups of data). Hence in the example in which $s = 4$ and $k - 1 = 2$ there are only two nonzero eigenvalues λ_1 and λ_2, and they are shown in the table. Only the two eigenvectors corresponding to the two largest eigenvalues of **D** are required for a two-dimensional ordination. (In the present case, of course, the two largest eigenvalues are the only nonzero eigenvalues.) These eigenvectors are shown as the rows of the $2 \times s = 2 \times 4$ matrix $\mathbf{W}'_{(2)}$.

7. The required coordinates for the data points are given by the columns of the $2 \times n = 2 \times 14$ matrix **Y**, defined as

$$\mathbf{Y} = \mathbf{W}'_{(2)}\mathbf{X}_{(4)},$$

in which $\mathbf{X}_{(4)}$ is the $s \times n = 4 \times 14$ matrix obtained by deleting the $k - 1 = 2$ rows of dummy variables from the centered and standardized data matrix.

The ordination is shown in Figure 6.3*a*. For comparison, a PCA ordination of the same data is given in Figure 6.3*b*.

As may be seen, the three sets of points are much more clearly differentiated by discriminant ordination than by PCA. The only outlier in the discriminant ordination is the northern Alberta station that seems to belong with the Yukon group of stations more than with its own group. This outlier is Fort Chipewyan; as may be seen in Table 6.1, it has the coldest winters and driest summers of the northern Alberta group.

The advantage of discriminant ordination is that it permits the differences among data sets to be displayed with maximum clarity. The process finds new coordinates (i.e., transformed scores) for several batches of data points in a way that ensures that each batch shall be as compact, and as widely separated from the other batches, as possible.

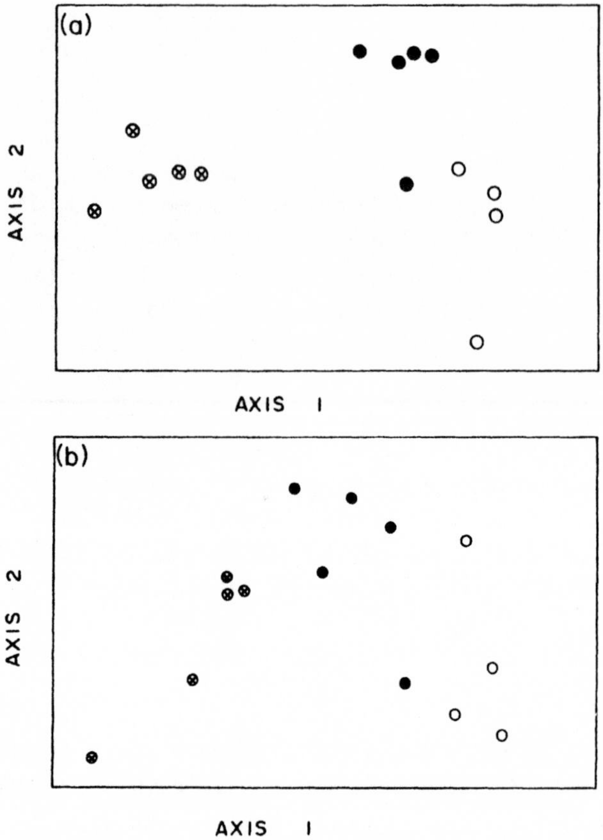

Figure 6.3. Two ordinations of the 14 weather stations: (*a*) discriminant ordination; (*b*) PCA ordination (centered and standardized). The symbols for the three regions are the same as in Figure 6.2.

The number of ways that now exist for classifying and ordinating ecological data is already large. No doubt the invention of new, more ingenious techniques will continue. It could be argued that the time has come for calling a halt to the endless proliferation of new methods. If ecologists are ever to fit their individual contributions together into a single unified body of scientific knowledge, it seems desirable that a few good methods of data analysis should be adopted widely and used consistently and that unproven methods should be consigned to the scrap-heap.

Attractive though this argument may be, it collapses before a more persuasive counterargument. This is that the development of data interpreting methods is an integral part of scientific progress and, as such, will not and should not come to a halt. Every improvement in computer capabilities will be followed by matching improvements in techniques for data interpretation, and if the interpretation of ecological data is to remain in the hands and minds of ecologists, they must familiarize themselves thoroughly with the general principles underlying methods of data handling.

EXERCISES

6.1. The following three 3 × 3 matrices $\mathbf{H}_1$, $\mathbf{H}_2$, and $\mathbf{H}_3$ do not have inverses.

$$\mathbf{H}_1 = \begin{pmatrix} 1 & 1 & 1 \\ 2 & 2 & 2 \\ 3 & 3 & 3 \end{pmatrix}; \quad \mathbf{H}_2 = \begin{pmatrix} 1 & 2 & 3 \\ 1 & 0 & -1 \\ -1 & -2 & -3 \end{pmatrix};$$

$$\mathbf{H}_3 = \begin{pmatrix} 1 & 1 & 1 \\ 1 & 2 & -2 \\ -3 & -6 & 6 \end{pmatrix}.$$

Use each in turn to multiply the 3 × 8 matrix $\mathbf{X}$ (which represents a cube) where

$$\mathbf{X} = \begin{pmatrix} 1 & 2 & 1 & 2 & 1 & 2 & 1 & 2 \\ 1 & 1 & 2 & 2 & 1 & 1 & 2 & 2 \\ 1 & 1 & 1 & 1 & 2 & 2 & 2 & 2 \end{pmatrix}.$$

Examine the products $\mathbf{H}_1\mathbf{X}$, $\mathbf{H}_2\mathbf{X}$, and $\mathbf{H}_3\mathbf{X}$ and determine the dimensionality of the figure into which the cube has been transformed in each case. What relationship does this suggest between matrices that cannot be inverted and the transformations that such matrices bring about?

6.2. Construct a diagram like that in Figure 3.4 (page 98) but with the y_1 and y_2-axes *not* perpendicular to each other. Let the angles between the old and new axes θ_{11}, θ_{12}, θ_{21}, and θ_{22} be defined as on page 100. Derive equations analogous to (3.4a) and (3.4b) on page 99.

6.3. Suppose the 6×3 matrix $\mathbf{X}$ is partitioned as shown into two submatrices denoted by $\mathbf{X}_1$ and $\mathbf{X}_2$.

$$\mathbf{X} = \begin{pmatrix} x_{11} & x_{12} & x_{13} \\ x_{21} & x_{22} & x_{23} \\ \hdashline x_{31} & x_{32} & x_{33} \\ x_{41} & x_{42} & x_{43} \\ x_{51} & x_{52} & x_{53} \\ x_{61} & x_{62} & x_{63} \end{pmatrix} = \begin{pmatrix} \mathbf{X}_1 \\ \hdashline \mathbf{X}_2 \end{pmatrix}$$

Write out the 6×6 SSCP matrix $\mathbf{XX}'$ in full, and show how it can be partitioned into four submatrices that are identical with those in the product

$$\begin{pmatrix} \mathbf{X}_1 \\ \hdashline \mathbf{X}_2 \end{pmatrix} \left(\mathbf{X}_1' \,\vdots\, \mathbf{X}_2' \right) = \begin{pmatrix} \mathbf{X}_1\mathbf{X}_1' & \vdots & \mathbf{X}_1\mathbf{X}_2' \\ \hdashline \mathbf{X}_2\mathbf{X}_1' & \vdots & \mathbf{X}_2\mathbf{X}_2' \end{pmatrix}.$$

(Keep track of the sizes of the various submatrices and their products.) Note that the multiplication of partitioned matrices is carried out according to the same rules as ordinary matrix multiplication except that submatrices take the place of individual elements.

Answers to Exercises

CHAPTER 2

2.1. (a) 10.15; (b) 8.49; (c) 10.10.

2.2.

Step	Fusion	"Farthest Points"	Distance Between Clusters
1	1, 3	1, 3	4.58
2	[1, 3], 2	2, 3	8.49
3	4, 5	4, 5	9.17

2.3. The coordinates of the centroids are:

Cluster [1, 2]	Cluster [3, 4, 5]
5.5	0.333
0	1.667
2	2.667
−3	1.667

The distance between them is 7.190.

2.4. $d^2([P],[M, N]) = 554.125$. (Note: The result is independent of p.)

2.5. 173.2.

2.6. (a) 0.2705; (b) 1.297; (c) 1.297 radians = 74.3°.

2.7. (a) $J = \frac{1}{3}$, $S = \frac{1}{2}$; (b) $J = \frac{3}{4}$, $S = \frac{6}{7}$; (c) $J = S = 1$.
Proof that $S \geq J$:

$$S = 2a/(2a + b + c) = 2a/(2a + f) \text{ on putting } b + c = f;$$

$$J = a/(a + f).$$

$$S/J = (2a + 2f)/(2a + f) > 1 \quad \text{when } f > 0 \quad \text{or} \quad = 1 \text{ when } f = 0.$$

CHAPTER 3

3.1.

(a) $\begin{pmatrix} 8 & -1 & 4 & 3 \\ 14 & 3 & 0 & 3 \end{pmatrix}$; (b) $\begin{pmatrix} 11 & 19 \\ -7 & -4 \\ 12 & 22 \end{pmatrix}$; (c) cannot be formed;

(d) $\begin{pmatrix} 3 & 0 & -1 \\ 5 & 4 & 1 \\ 5 & -2 & -3 \\ 23 & 10 & -1 \end{pmatrix}$; (e) cannot be formed;

(f) $\begin{pmatrix} 68 & 22 & -8 \\ -19 & -14 & -3 \\ 78 & 24 & -10 \end{pmatrix}$ (g) $\begin{pmatrix} 14 & 3 & 0 & 3 \\ 30 & 1 & 8 & 9 \\ 20 & 7 & -4 & 3 \\ 124 & 13 & 20 & 33 \end{pmatrix}$.

[Note: To form the product **BCA**, for example, one may postmultiply **BC** by **A** or **B** by **CA**.]

3.2.

$$\mathbf{U}_1 = \begin{pmatrix} 2 & -1 \\ 1 & 3 \end{pmatrix}; \quad \mathbf{U}_2 = \begin{pmatrix} -1 & -1 \\ -2 & 2 \end{pmatrix}; \quad \mathbf{U}_3 = \begin{pmatrix} 0 & 2 \\ -3 & 1 \end{pmatrix}.$$

3.3. $\mathbf{U}_2$ is orthogonal; $\mathbf{U}_1$ is not.

3.4. Let $\mathbf{XX}' = \mathbf{A}$. Then a_{ij} is the sum of cross-products of the ith row of **X** and the jth column of $\mathbf{X}'$ (which is the jth row of **X**). Likewise, a_{ji} is the sum of cross-products of the jth row of **X** and the ith column of $\mathbf{X}'$ (which is the ith row of **X**). Hence $a_{ij} = a_{ji}$ for all i, j.

3.5.

$$\begin{pmatrix} 1 & -0.5 \\ -0.5 & 1 \end{pmatrix}.$$

3.6. The eigenvalues of $\mathbf{A}^5$ are $2^5 = 32$ and $3^5 = 243$. This follows from:

$$\begin{aligned} \mathbf{A}^5 &= (\mathbf{U}'\Lambda\mathbf{U})(\mathbf{U}'\Lambda\mathbf{U})(\mathbf{U}'\Lambda\mathbf{U})(\mathbf{U}'\Lambda\mathbf{U})(\mathbf{U}'\Lambda\mathbf{U}) \\ &= \mathbf{U}'\Lambda(\mathbf{U}\mathbf{U}')\Lambda(\mathbf{U}\mathbf{U}')\Lambda(\mathbf{U}\mathbf{U}')\Lambda(\mathbf{U}\mathbf{U}')\Lambda\mathbf{U} \\ &= \mathbf{U}'\Lambda^5\mathbf{U} \quad \text{since } \mathbf{U}\mathbf{U}' = \mathbf{I}. \end{aligned}$$

Hence the eigenvalues of $\mathbf{A}^5$ are the eigenvalues of $\mathbf{A}$ raised to the fifth power.

3.7. $(0.6733 \quad 0.5858 \quad 0.4242 \quad -0.1347 \quad -0.0741)$.

3.8. $\lambda_2 = 2.55$ [from $\text{tr}(\mathbf{S}) = \Sigma_i \lambda_i$].

3.10. λ_2 is the same for **G** and **F**. Hence $\lambda_2 = 45.285$. The second eigenvector of **F** is

$$\mathbf{u}_2' = (u_{21} \quad u_{22}) = (-u_{12} \quad u_{11}) = (-0.807 \quad 0.590).$$

Then the second eigenvector of **G**, namely, $\mathbf{v}_2'$, is proportional to $\mathbf{u}_2'\mathbf{X}$. Hence $\mathbf{v}_2' = (0.61 \quad -0.78 \quad -0.14)$.

CHAPTER 4

4.1. The covariance matrix is $(1/n)\mathbf{R}$ with $n = 8$. Hence $\lambda_1 = 36$; $\lambda_2 = 25$; $\lambda_3 = 16$.

4.2.

$\left.\begin{matrix}\text{Tr}(\mathbf{XX}') \\ \text{Tr}(\mathbf{YY}')\end{matrix}\right\}$ is the sum of squares of the distances of all points in the swarm from the origin of the $\left\{\begin{matrix} x\text{-coordinate frame} \\ y\text{-coordinate frame}\end{matrix}\right.$

Since the origins of the two coordinate frames coincide at the centroid of the swarm, $\text{tr}(\mathbf{XX}') = \text{tr}(\mathbf{YY}')$. [From this it follows easily that $(1/n)\text{tr}(\mathbf{XX}') = \Sigma_i \lambda_i$; see page 126.]

4.3. Let the 2×2 correlation matrix be

$$\mathbf{P} = \begin{pmatrix} 1 & \rho \\ \rho & 1 \end{pmatrix}.$$

Let the matrix of eigenvectors be

$$\mathbf{U} = \begin{pmatrix} \cos\theta & \sin\theta \\ -\sin\theta & \cos\theta \end{pmatrix}.$$

Then since $\mathbf{UPU}' = \Lambda$, it follows that the (1, 2)th element of $\mathbf{UPU}'$ is 0. That is, $\rho(\cos^2\theta - \sin^2\theta) = 0$. Hence

$$\cos^2\theta = \sin^2\theta; \qquad \cos\theta = \pm\sin\theta; \qquad \theta = 45°.$$

Therefore the eigenvectors are:

$$(\cos\theta \quad \sin\theta) = (0.7071 \quad 0.7071)$$

and

$$(-\sin\theta \quad \cos\theta) = (-0.7071 \quad 0.7071).$$

4.4. (a) 10.6°; (b) 25.0°; (c) 55.9°; (d) 45°.

4.5. 0.816.

4.7. The ordinated data form a triangle the lengths of whose sides are:

$$d(\mathrm{A,B}) = 4; \qquad d(\mathrm{B,C}) = 3; \qquad d(\mathrm{A,C}) = 7.$$

4.9. The following equations are numbered to correspond with those in the text, except that primes have been added here.

$$\mathbf{w} = \mathbf{C}^{-1}\mathbf{X}'\mathbf{R}^{-1}\mathbf{X}\mathbf{w} \qquad (4.22')$$

$$\mathbf{C}^{1/2}\mathbf{w} = \mathbf{C}^{-1/2}(\mathbf{X}'\mathbf{R}^{-1}\mathbf{X})(\mathbf{C}^{-1/2}\mathbf{C}^{1/2})\mathbf{w} \qquad (4.24')$$

$$= (\mathbf{C}^{-1/2}\mathbf{X}'\mathbf{R}^{-1/2})(\mathbf{R}^{-1/2}\mathbf{X}\mathbf{C}^{-1/2})(\mathbf{C}^{1/2}\mathbf{w}) \qquad (4.25')$$

$$= \mathbf{Q}(\mathbf{C}^{1/2}\mathbf{w}). \qquad (4.26')$$

Therefore,

$$\mathbf{w}'\mathbf{C}^{1/2} = \mathbf{w}'\mathbf{C}^{1/2}\mathbf{Q} \qquad (4.27')$$

whence

$$\mathbf{W}\mathbf{C}^{1/2} \propto \mathbf{U}_{\mathbf{Q}} \qquad (4.28')$$

where $\mathbf{U}_{\mathbf{Q}}$ is the $n \times n$ matrix of eigenvectors of **Q**. Therefore,

$$\mathbf{W} \propto \mathbf{U}_{\mathbf{Q}}\mathbf{C}^{-1/2}. \qquad (4.29')$$

4.10.

$$\mathbf{Y} = \begin{pmatrix} -12.99 & -4.33 & 4.33 & 12.99 \\ 0 & 0 & 0 & 0 \\ 0 & 0 & 0 & 0 \\ 0 & 0 & 0 & 0 \\ 0 & 0 & 0 & 0 \end{pmatrix}.$$

This follows from the fact that the data points lie on a straight line in five-space (i.e., they are confined to a one-dimensional subspace of the five-space). The PCA places the first principal axis on this line; therefore, the coordinates of the points can be found by determining the distances separating them. An eigenanalysis of the covariance matrix would show that it has only one nonzero eigenvalue. The number of nonzero eigenvalues of a covariance matrix, which is equal to the number of dimensions of the subspace in which the data points lie, is known as the *rank* of the covariance matrix. In this example, the rank is 1.

CHAPTER 5

5.1. The segments of the minimum spanning tree are:

1: $d(\mathrm{C},\mathrm{E}) = 1.54$; 2: $d(\mathrm{A},\mathrm{E}) = 1.74$; 3: $d(\mathrm{B},\mathrm{A}) = 1.88$;

4: $d(\mathrm{D},\mathrm{A}) = 2.26$; 5: $d(\mathrm{F},\mathrm{A}) = 2.93$; 6: $d(\mathrm{G},\mathrm{A}) = 3.30$;

7: $d(\mathrm{I},\mathrm{D}) = 8.20$; 8: $d(\mathrm{J},\mathrm{I}) = 3.00$; 9: $d(\mathrm{H},\mathrm{I}) = 3.27$.

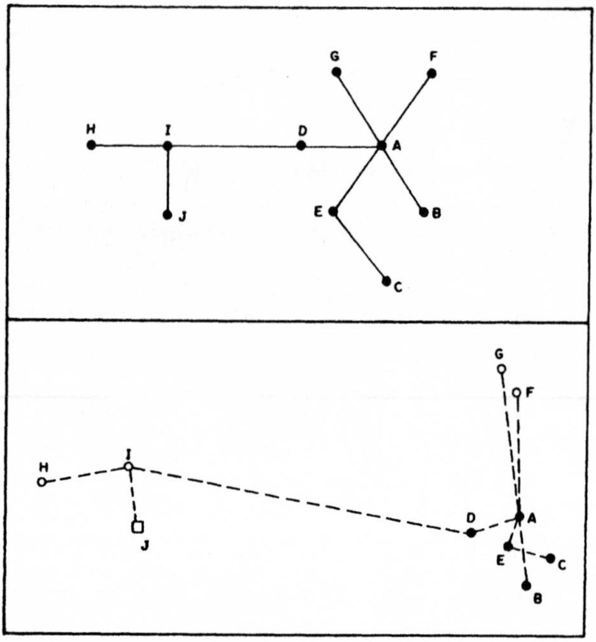

One of the many possible diagrammatic representations of the minimum spanning tree is shown in the figure (upper panel). Compare it with the two-dimensional ordination of the data, with the minimum spanning tree superimposed, in the lower panel. The lower panel is a reproduction of Figure 5.3 (page 210) with the points labeled to match. The distance matrix in the exercise was taken from Gower and Ross (1969).

5.2. Successive divisions give the following classes:

1: (A, B) (C, D, E, F);

2: (A) (B) (C, D, E) (F);

3: (A) (B) (C) (D, E) (F);

4: (A) (B) (C) (D) (E) (F).

Notice that A and B are separated at the second step even though they still form a close pair in the ordination pattern which is almost indistinguishable from the pattern in Figure 5.4*a*. This "unnatural" separation occurs because the second principal axis passes between A and B.

5.3. Successive divisions give the following classes:

1: (A, B) (C, D, E, F);

2: (A, B) (C, D, E) (F);

3: (A, B) (C) (D, E) (F).

Observe that the close pair (A, B) has not been divided.

CHAPTER 6

6.1. The points of $\mathbf{H}_1\mathbf{X}$ form a straight line, hence form a figure of one dimension. The points of $\mathbf{H}_2\mathbf{X}$ (and also of $\mathbf{H}_3\mathbf{X}$) are confined to a plane, and hence form a figure of two dimensions. This leads to the conjecture that a matrix can be inverted only if it brings about no reduction in the dimensionality of a swarm of points when it is used to transform the swarm. Proof of the correctness of this conjecture, and methods for judging whether a given matrix can be inverted, are beyond the scope of this book. See, for example, Searle (1966), Chapter 4, or Tatsuoka (1971), Chapter 5. Matrices that can, and cannot, be inverted are called, respectively, "nonsingular" and "singular."

6.2. The equations are

$$y_1 = x_1\cos\theta_{11} + x_2\cos\theta_{12}$$

$$y_2 = x_1\cos\theta_{21} + x_2\cos\theta_{22}.$$

Glossary

(Words in italics are defined elsewhere in the glossary.)

Agglomerative classification. Same as *clustering.*

Arch effect. The appearance of a projected *data swarm* as a curve ("arch") when the data were obtained from *sampling units* ranged along a one-dimensional gradient.

Asymmetry, coefficient of. A measure of the degree to which an *ordination* axis approaches unipolarity (see *unipolar axis*).

Average distance between two clusters. The arithmetic average of all the distances between a point in one cluster and a point in the other.

Average linkage clustering. Collective term for all clustering methods in which the distance between two clusters depends on the locations of all points in both clusters. Contrast *nearest-neighbor* and *farthest-neighbor clustering.*

Average linkage clustering criterion. The dissimilarity between clusters [P] and [Q], where [Q] is formed by the fusion of clusters [M] and [N]. Four measures of this dissimilarity are:

Centroid distance, from the centroid of [P] to the centroid of [Q].

Median distance, from the centroid of [P] to the midpoint of the line joining the centroids of [M] and [N].

Unweighted average distance, same as the *average distance* between [P] and [Q].

Weighted average distance, the average of the average distance between [P] and [M] and the average distance between [P] and [N].

Between-axes heterogeneity. The heterogeneity exhibited by a *data swarm* consisting of two or more subswarms confined (or almost confined) to different subspaces of the total space containing the whole swarm. Contrast *within-axes heterogeneity.*

Binary data. Data consisting entirely of zeros and ones.

$$x_{ij} = \begin{cases} 1 & \text{if species } i \text{ is present in } quadrat\ j, \\ 0 & \text{otherwise.} \end{cases}$$

Bipolar axis. An *ordination* axis on which the *data points* have some positive and some negative scores. Contrast *unipolar axis.*

Catenation. An *ordination* designed to show, as clearly as possible, the structure of *nonlinear data.*

Centered data. Data in which the observations are expressed as deviations from their mean value. Hence their sum is zero.

Centroid. The center of gravity (or "average point") of a swarm (or cluster) of points in a space of any number of dimensions. The coordinate of the centroid on each axis is the mean of the coordinates on that axis of all the points in the swarm (or cluster).

Centroid clustering. A clustering technique in which the distance (or dissimilarity) between two clusters is set equal to the distance (or dissimilarity) between their *centroids.*

Centroid distance. See *average linkage clustering criterion.*

Chaining. In a *clustering* process, the tendency for one cluster to grow by the repeated addition, one at a time, of single points.

Characteristic values or **roots.** Same as *eigenvalues.* See *eigenanalysis.*

Characteristic vector. Same as *eigenvector.* See *eigenanalysis.*

Chord distance. The shortest (straight line) distance between two points on the same circle, sphere, or hypersphere.

City-block distance, CD. The distance between two points, in a coordinate frame of any number of dimensions, measured as the sum of segments parallel with the axes. For points j and k,

$$\text{CD} = \sum_{i=1}^{s} |x_{ij} - x_{ik}|.$$

Clustering. The process of classifying *data points* by combining similar points to form small classes, then combining small classes into larger classes, and so on. Same as *agglomerative classification*. Contrast *divisive classification*.

Column vector. A *matrix* with only one column. Equivalently, a *vector* written as a vertical column.

Complete-linkage clustering. Same as *farthest-neighbor clustering.*

Combinatorial clustering methods. Those in which each successive *distance matrix* can be constructed from the preceding distance matrix; the raw data are needed only to construct the first distance matrix.

Correlation coefficient. A standardized form of *covariance* obtained by dividing the covariance of two variables, say x and y, by the product of the standard deviations of x and y. Its value always lies in $[-1, 1]$. It measures the degree to which x and y are related.

Correlation matrix. A *symmetric matrix* in which the (h, i)th *element*, when $h \neq i$, is the *correlation coefficient* between the hth and ith variables. All the elements on the main diagonal (top left to bottom right) are 1.

Covariance. When two variables, say x and y, are measured on each of a number of *sampling units*, their covariance is the mean of the cross-products of the *centered data*. The ith cross-product is $(x_i - \bar{x})(y_i - \bar{y})$ where $\bar{x}$ and $\bar{y}$ are the means of the xs and ys.

Covariance matrix. A *symmetric matrix* in which the (i, i)th *element* is the *variance* of the ith variable, and the (h, i)th element is the *covariance* of the hth and ith variables.

Czekanowski's Index of Similarity. Same as *percentage similarity*.

Data matrix. A numerical table in which each column lists all the observations on one *sampling unit* (or *quadrat*), and each row lists the values of one of the observed variables in all quadrats.

Data point. A geometric representation in multidimensional space of one column of a *data matrix*.

Data swarm. The set of all *data points* which usually occupies a space of many dimensions.

Dendrogram. A diagram showing the hierarchical relationships produced by a *hierarchical classification*.

Diagonal matrix. A square *matrix* in which all *elements* except those on the main diagonal (top left to bottom right) are zero.

Direction cosines. The cosines of the angles made by any line through the origin of a coordinate frame and the axes of the frame. Equivalently, the lengths of the projections onto the axes of a unit segment of the line.

Distance matrix. A *matrix* showing the distance from each point to every other point in a *data swarm*.

Divisive classification. The process of classifying *data points* by first dividing the whole swarm of points into classes, then redividing some or all of these classes into subclasses, and so on. Contrast *clustering*.

Eigenanalysis. The process of finding the *eigenvalue–eigenvector pairs* of a square *matrix* **A**. The eigenvalues are the elements of the diagonal matrix Λ and the eigenvectors are the rows of **U** (equivalently, the columns of **U**′) where $\mathbf{A} = \mathbf{U}'\Lambda\mathbf{U}$.

Eigenvalue–eigenvector pair of a *matrix* **A**. An eigenvalue (or "eigenscalar")–eigenvector pair of **A** are, respectively, a scalar number λ and a row vector $\mathbf{u}'$ related by the equation $\mathbf{u}'\mathbf{A} = \lambda\mathbf{u}'$. If **A** is an $n \times n$ matrix, there are n such pairs. See also *eigenanalysis*.

Element of a *matrix*. One of the individual numbers composing a matrix. The (i, j)th element is the number in the ith row and the jth column of the matrix.

Euclidean distance. The distance between two points in the ordinary sense in one, two, or three dimensions, or the conceptual analogue of distance in spaces of more than three dimensions.

Farthest-neighbor clustering. *Clustering* in which the distance (dissimilarity) between two clusters is taken to be the longest distance between a pair of points with one member of the pair in each cluster. Contrast *nearest-neighbor clustering*.

Geodesic metric. The great circle distance (shortest over-the-surface distance) between two points on a sphere or hypersphere.

Group-average clustering methods. *Clustering* methods that use the *unweighted* or *weighted average distance* as measures of the dissimilarity of a pair of clusters.

Hierarchical classification. A classification in which the classes are ranked. Every individual belongs to a class, and every class to a higher-ranking class, up to the highest class which is the totality of all individuals.

Horseshoe effect. Same as *arch effect.*

Identity matrix. A square *matrix* in which all the *elements* on the main diagonal (top left to bottom right) are ones and all other elements are zeros.

Internode. See *node.*

Inverse of a square matrix. The inverse of an $n \times n$ *matrix* $\mathbf{X}$ is the $n \times n$ matrix $\mathbf{X}^{-1}$ such that $\mathbf{XX}^{-1} = \mathbf{X}^{-1}\mathbf{X} = \mathbf{I}$. ($\mathbf{I}$ is the *identity matrix.*)

Jaccard's Index of Similarity between two quadrats. The ratio $a/(a + f)$ where a is the number of species common to both quadrats and f is the number of species present in one or other (but not both) of the quadrats. Compare *Sørensen's Index.*

Latent values or **roots.** Same as eigenvalues. See *eigenvalue–eigenvector pair.*

Latent vector. Same as eigenvector. See *eigenvalue–eigenvector pair.*

Linear data. A data swarm is (approximately) linear if its projection onto any two-dimensional space, however oriented, gives a two-dimensional swarm whose long axis is (approximately) a straight line. If any projection yields a (projected) swarm with a curved axis, then the data are *nonlinear.*

Linear transformation. A transformation of one set of points into another done by defining the coordinates of the transformed points as linear functions of their coordinates before transformation. The original coordinates appear only in the first degree (i.e., they are never squared or raised to a higher power).

Manhattan metric. Same as *city-block distance.*

Marczewski–Steinhaus distance, MS. A measure of the dissimilarity of two quadrats, the complement of *Jaccard's Index of Similarity.* $\text{MS} = f/(a + f)$ where f is the number of species present in only one (not both) of the quadrats, and a is the number of species common to both of them.

Matrix. A two-dimensional array of numbers. The meaning of each number (or *element*) depends on its position in the matrix, that is, on the row and the column in which it occurs.

Matrix multiplication. The formation of the product $\mathbf{AB}$ of two matrices $\mathbf{A}$ and $\mathbf{B}$. The (i, j)th *element* of $\mathbf{AB}$ is the sum of the pairwise products of the elements in the ith row of **A** and the jth column of **B**. Hence **AB** exists only if the number of columns in **A** is equal to the number of rows in **B**; **AB** has the same number of rows as **A** and the same number of columns as **B**; **AB** is **A** *postmultiplied* by **B** or, equivalently, **B** *premultiplied* by **A**. In general, $\mathbf{AB} \neq \mathbf{BA}$.

Median clustering method. A *clustering* method that uses the *median distance* between clusters as a dissimilarity measure.

Median distance. See *average-linkage clustering criterion*.

Metric measures of dissimilarity. Measures that, like distance, satisfy the *triangle inequality axiom*.

Minimum spanning tree. The shortest *spanning tree* that can be constructed in a given swarm of points.

Minimum variance clustering. *Clustering* in which the two clusters united at each step are those whose fusion brings about the smallest possible increase in *within-cluster dispersion*.

Monotonic clustering methods. Methods in which the occurrence of *reversals* is impossible.

Nearest-neighbor clustering. *Clustering* in which the distance (dissimilarity) between two clusters is taken to be the shortest distance between a pair of points with one member of the pair in each cluster. Contrast *farthest-neighbor clustering*.

Nodes and **internodes.** The parts of a *dendrogram*. The nodes are the horizontal lines linking classes of equal rank. The internodes are the vertical lines linking each class to the classes above and below it in rank.

Nonlinear data. See *linear data*.

Normalized data. The coordinates of a data point or the elements of a vector rescaled so that their squares sum to unity.

Ordination. The ordering of a set of *data points* with respect to one or more axes. Alternatively, the displaying of a swarm of data points in a two or three-dimensional coordinate frame so as to make the relationships among the points in many-dimensional space visible on inspection.

Ordination-space partitioning. The placing of partitions in an ordinated swarm of *data points* in order to separate the points into groups or classes. The result is a *divisive classification*.

Orthogonal matrix. A square *matrix* that, when used as a *transformation matrix*, causes a rigid rotation of the *data swarm* around the origin of the coordinate frame without any change of scale. The product of an orthogonal matrix and its *transpose* is the *identity matrix*.

Partitioned matrix. A *matrix* that has been subdivided into *submatrices* by placing one or more horizontal (between-row) partitions and/or one or more vertical (between-column) partitions so as to divide the matrix into rectangular blocks.

Percentage difference. Same as *percentage dissimilarity.*

Percentage dissimilarity, PD. The complement of *percentage similarity* PS. PD = 100 − PS.

Percentage distance. Same as *percentage dissimilarity.*

Percentage remoteness, PR [new term]. The complement of *Ružička's Index of Similarity* RI. PR = 100 − RI.

Percentage similarity. The percentage similarity of *quadrats* j and k is

$$S = 200 \sum_{i=1}^{s} \frac{\min(x_{ij}, x_{ik})}{x_{ij} + x_{ik}} \quad \text{percent}$$

where x_{ij} and x_{ik} are the quantities of species i in quadrats j and k, and $\min(x_{ij}, x_{ik})$ is the lesser of the two quantities.

Pool [in this book]. A batch of replicate *quadrats* from a homogeneous population. Differences among them are due only to chance.

Postmultiply. See *matrix multiplication.*

Premultiply. See *matrix multiplication.*

Principal axes. The new coordinate axes for a swarm of *data points*, obtained by doing a principal component analysis of the data. Each axis represents a *principal component* of the data.

Principal components. New variables derived by a principal component analysis to describe a body of data. Each is a weighted sum of the "raw" (as originally measured) variables, or of the *centered* and/or *standardized* variables.

Principal component score. The value of a *principal component* for an individual point. Hence the coordinate of the point on the corresponding *principal axis.*

Q-type ordination. An *ordination* of species. The *data points* represent species and the coordinate axes (before ordination) represent *quadrats.* The (j, k)th *element* of the *covariance matrix* analyzed is the *covariance* of the quantities (of all species) in quadrats j and k. Contrast *R-type ordination.*

Quadrat. In this book, an ecological *sampling unit* of any kind.

R-type ordination. An ordination of *quadrats.* The more usual form of ordination. The *data points* represent quadrats and the coordinate axes (before ordination) represent species. The (h, i)th *element* of the *covariance matrix* analyzed is the *covariance* of the quantities (in all quadrats) of species h and i. Contrast *Q-type ordination.*

Rank of a *covariance matrix.* The number of its nonzero *eigenvalues.* Equivalently, the number of dimensions of the space in which the *data points* lie.

Residual matrix. The rth residual matrix of the square *symmetric matrix* $\mathbf{A}$ is

$$\mathbf{A}_r = \mathbf{A} - \lambda_1 \mathbf{u}_1 \mathbf{u}_1' - \lambda_2 \mathbf{u}_2 \mathbf{u}_2' - \cdots - \lambda_r \mathbf{u}_r \mathbf{u}_r'$$

where λ_i and $\mathbf{u}_i$ are the ith eigenvalue and eigenvector of $\mathbf{A}$.

Reversal (in *clustering*). A reversal occurs when a fusion made late in a clustering process unites clusters that are closer together than were the clusters joined at an earlier fusion.

Row vector. A *matrix* with only one row. Equivalently, a *vector* written as a horizontal row.

Ružička's Index of Similarity, RI between quadrats j and k is

$$\text{RI} = 100 \sum_{i=1}^{s} \frac{\min(x_{ij}, x_{ik})}{\max(x_{ij}, x_{ik})} \quad \text{percent}$$

where x_{ij} and x_{ik} are the quantities of species i in quadrats j and k; $\min(x_{ij}, x_{ik})$ and $\max(x_{ij}, x_{ik})$ denote, respectively, the lesser and the greater of the two quantities.

Sample. A collection of sampling units or *quadrats.*

Sampling unit. An individual plot or *quadrat.* A collection of many such units, each of which is a different small fragment of the community under study, constitutes a *sample.*

Scalar. An "ordinary" number, in contrast to an array of numbers (a *matrix*).

Single-linkage clustering. Same as *nearest-neighbor clustering*.

Sørensen's Index of Similarity between two *quadrats*. The ratio $2a/(2a + f)$,where a is the number of species common to both quadrats and f is the number of species present in one or other (but not both) of the quadrats. Compare *Jaccard's Index of Similarity*.

Spanning tree. A set of line segments joining all the points in a swarm of points so that every pair of points is linked by only one path (i.e., there are no loops).

SSCP matrix. Same as *sum-of-squares-and-cross-products matrix*.

Standard deviation. The square root of the *variance*.

Standardized data. Data that have been rescaled by dividing every observation by the *standard deviation* of all the observations.

Stopping rule. A rule for deciding when a *divisive classification* should stop.

Submatrix. A subset of a given *matrix* that is itself a matrix. It is delimited en bloc from the "parent" matrix, with the arrangement of the *elements* unchanged. See also *partitioned matrix*.

Sum-of-squares-and-cross-products matrix. The *matrix* obtained by multiplying a *data matrix* by its *transpose*. The (i, i)th *element* is the sum of squares of the ith variable. The (h, i)th element is the sum of cross-products of the hth and ith variables.

Symmetric matrix. A square *matrix* that is symmetric about its main diagonal (top left to bottom right). Thus the (h, i)th *element* and the (i, h)th element are equal for all h, i. A square matrix of which this is not true is *unsymmetric*.

Trace of a square *matrix*. The sum of the *elements* on the main diagonal (top left to bottom right). The trace of $\mathbf{A}$ is written $\mathrm{tr}(\mathbf{A})$.

Transformation matrix. A *matrix* used to *premultiply* a *data matrix* in order to bring about a *linear transformation* of the data.

Transpose of a matrix. The transpose of the $s \times n$ *matrix* $\mathbf{X}$ is the $n \times s$ matrix having the rows of $\mathbf{X}$ as its columns (and, consequently, the columns of $\mathbf{X}$ as its rows). It is denoted by $\mathbf{X}'$.

Tree diagram. Same as *dendrogram*.

Triangle inequality axiom. The axiom that the distance between any two points A and B cannot exceed the sum of the distances from each of them to a third point C; that is, $d(\mathrm{A}, \mathrm{B}) \leq d(\mathrm{A}, \mathrm{C}) + d(\mathrm{B}, \mathrm{C})$.

Ultrametric distance measures. Those that cannot in any circumstances cause a *reversal* in a *clustering* process.

Unipolar axis. An *ordination* axis on which the data points all have scores of the same sign (all positive or all negative). Contrast *bipolar axis*.

Unsymmetric matrix. See *symmetric matrix*.

Unweighted average distance. Same as *average distance*. See *average-linkage clustering criteria*.

Variance. The mean of the squared deviations, from their mean value, of a set of observations.

Variance-covariance matrix. Same as *covariance matrix*.

Vector. A *row vector* or *column vector*. In some contexts, the n (say) *elements* of a vector constitute the coordinates of a point in n-dimensional space.

Weighted average distance. See *average-linkage clustering criteria*.

Weighted centroid distance. Same as *median distance*. See *average-linkage clustering criteria*,

Weighted and unweighted clustering methods. Weighted methods treat clusters as of equal weight irrespective of their numbers of points. Unweighted methods treat *data points* as of equal weight so that the weight of a cluster is proportional to its number of points.

Within-axes heterogeneity. The heterogeneity exhibited by a *data swarm* consisting of two or more subswarms, when all subswarms occupy the same many-dimensional space. Contrast *between-axes heterogeneity*.

Within-cluster dispersion of a cluster. The sum of squares of the distances from every point in the cluster to the cluster's centroid.

Bibliography

Anderberg, M. R. (1973). *Cluster Analysis for Applications*. Academic Press, New York.

Carleton, T. J., and P. F. Maycock (1980). Vegetation of the boreal forests south of James Bay: non-centered component analysis of the vascular flora. *Ecology* **61**: 1199–1212.

Delaney, M. J., and M. J. R. Healy (1966). Variation in the white-toothed shrews (*Crocidura* spp) in the British Isles. *Proc. Roy. Soc. B.* **164**: 63–74.

Gauch, H. G., Jr. (1979). *COMPCLUS—A FORTRAN program for rapid initial clustering of large data sets*. Cornell University, Ithaca, N.Y.

Gauch, H. G. (1980). Rapid initial clustering of large data sets. *Vegetatio* **42**: 103–111.

Gauch, H. G., Jr. (1982a). *Multivariate Analysis in Community Ecology*. Cambridge University Press.

Gauch, H. G., Jr. (1982b). Noise reduction by eigenvector ordinations. *Ecology* **63**: 1643–1649.

Gauch, H. G., Jr., and R. H. Whittaker (1972). Coenocline simulation. *Ecology* **53**: 446–451.

Gauch, H. G., Jr., and R. H. Whittaker (1981). Hierarchical classification of community data. *J. Ecol.* **69**: 135–152.

Goodall, D. W. (1978a). Sample similarity and species correlation. In "Ordination of Plant Communities" (R. H. Whittaker, Ed.), W. Junk, The Hague, pp. 99–149.

Goodall, D. W. (1978b). Numerical classification. In "Classification of Plant Communities" (R. H. Whittaker, Ed.), W. Junk, The Hague, pp. 249–288.

Gordon, A. D. (1981). *Classification. Methods for the Exploratory Analysis of Multivariate Data*. Chapman and Hall, New York.

Gower, J. C. (1967). A comparison of some methods of cluster analysis. *Biometrics* **23**: 623–637.

Gower, J. C., and P. G. N. Digby (1981). Expressing complex relationships in two dimensions. In "Interpreting Multivariate Data" (V. Barnett, Ed.), Wiley, New York, pp. 83–118.

Gower, J. C., and G. J. S. Ross (1969). Minimum spanning trees and single linkage cluster analysis. *Appl. Statist.* **18**: 54–64.

Hill, M. O. (1973). Reciprocal averaging: an eigenvector method of ordination. *J. Ecol.* **61**: 237–251.

Hill, M. O. (1979a). *DECORANA—A FORTRAN Program for Detrended Correspondence Analysis and Reciprocal Averaging*. Cornell University, Ithaca, NY.

Hill, M. O. (1979b). *TWINSPAN—A FORTRAN Program for Arranging Multivariate Data in an Ordered Two-Way Table by Classification of the Individuals and Attributes*. Cornell University, Ithaca, NY.

Hill, M. O., R. G. H. Bunce, and M. W. Shaw (1975). Indicator species analysis, a divisive polythetic method of classification, and its application to a survey of native pinewoods in Scotland. *J. Ecol.* **63**: 597–613.

Hill, M. O., and H. G. Gauch, Jr. (1980). Detrended correspondence analysis, an improved ordination technique. *Vegetatio* **42**: 47–58.

Jeglum, J. K., C. F. Wehrhahn, and M. A. Swan (1971). Comparisons of environmental ordinations with principal component vegetational ordinations for sets of data having different degrees of complexity. *Can. J. Forest Res.* **1**: 99–112.

Kempton, R. A. (1981). The stability of site ordinations in ecological surveys. In *The Mathematical Theory of the Dynamics of Biological Populations II* (R. W. Hiorns and D. Cooke, Eds). Academic Press, New York, pp. 217–230.

Lance, G. N., and W. T. Williams (1966). A general theory of classificatory sorting strategies. 1. Hierarchical systems. *Computer J.* **9**: 373–380.

Lefkovitch, L. P. (1976). Hierarchical clustering from principal coordinates: an efficient method for small to very large numbers of objects. *Math. Biosci.* **31**: 157–174.

Levandowsky, M. (1972). An ordination of phytoplankton populations in ponds of varying salinity and temperature. *Ecology* **53**: 398–407.

Levandowsky, M., and D. Winter (1971). Distance between sets. *Nature* **234**: 34–35.

Lieffers, V. J. (1984). Emergent plant communities of oxbow lakes in northeastern Alberta: Salinity, water level fluctuations and succession. *Can. J. Botany* **62**: 310–316.

Maarel, E. van der (1980). On the interpretability of ordination diagrams. *Vegetatio* **42**: 43–45.

Maarel, E. van der, J. G. M. Janssen, and J. M. W. Louppen (1978). TABORD, a program for structuring phytosociological tables. *Vegetatio* **38**: 143–156.

Marks, P. L., and P. A. Harcombe (1981). Forest vegetation of the Big Thicket, southeast Texas. *Ecol. Monogr.* **51**: 287–305.

Morrison, D. F. (1976). *Multivariate Statistical Methods*. 2nd ed. McGraw-Hill, New York.

Newnham, R. M. (1968). A classification of climate by principal component analysis and its relationship to tree species distribution. *Forest Sci.* **14**: 254–264.

Nichols, S. (1977). On the interpretation of principal component analysis in ecological contexts. *Vegetatio* **34**: 191–197.

Noy-Meir, I. (1973a). Data transformations in ecological ordination. I Some advantages of non-centering. *J. Ecol.* **61**: 329–341.

Noy-Meir, I. (1973b). Divisive polythetic classification of vegetation data by optimized divisions on ordination components. *J. Ecol.* **61**: 753–760.

Noy-Meir, I. (1974). Catenation: quantitative methods for the definition of coenoclines. *Vegetatio* **29**: 89–99.

Noy-Meir, I., D. Walker, and W. T. Williams (1975). Data transformations in ecological ordination. II. On the meaning of data standardization. *J. Ecol.* **63**: 779–800.

Orlóci, L. (1978). *Multivariate Analysis in Vegetation Research*. W. Junk, The Hague.

Pielou, E. C. (1977). *Mathematical Ecology*. Wiley, New York.

Rohlf, F. J. (1973). Algorithm 67: Hierarchical clustering using the minimum spanning tree. *Computer J.* **16**: 93–95.

Ross, G. J. S. (1969). Algorithm AS 13–15. *Appl. Statist.* **18**: 103–110.

Searle, S. R. (1966). *Matrix Algebra for the Biological Sciences*. Wiley, New York.

Shepard, R. N., and J. D. Carroll (1966). Parametric representations of non-linear data structures. In "Multivariate Analysis" (P. R. Krishnaiah, Ed.), Academic Press, New York.

Sneath, P. H. A., and R. R. Sokal (1973). *Numerical Taxonomy*. W. H. Freeman & Co., San Francisco.

Strauss, R. E. (1982). Statistical significance of species clusters in association analysis. *Ecology* **63**: 634–639.

Tatsuoka, M. M. (1971). *Multivariate Analysis*. Wiley, New York.

Whittaker, R. H. (Ed.) (1978a). *Ordination of Plant Communities*. W. Junk, The Hague.

Whittaker, R. H. (1978b). *Classification of Plant Communities*. W. Junk, The Hague.

Index

Page numbers in **boldface** indicate substantial treatment of a topic.

Printed in the United States
38933LVS00003B/5

9 780471 889502